# STELLINGEN

## I

De Erbstockhypothese van PLATE vindt weinig steun in de feiten.

## II

Het bestaan van *Leporiden* is niet exakt bewezen.

NACHTSHEIM, Z. f. Züchtung, 33, 1935

## III

Het is onwaarschijnlijk, dat parallelindukties ontstaan ten gevolge van een gelijktijdige induktie op soma en geslachtscellen.

## IV

Het is moeilijk bij kleureigenschappen, die op multiple allelen berusten, een verschillende kwantiteit der genen aan te nemen.

## V

De door SCHMALFUSS gegeven verklaring van dominantie verdient in bepaalde gevallen de voorkeur boven die van GOLDSCHMIDT.

## VI

Het is mogelijk de methode van BARGER tot het bepalen van de osmotische waarde van een kleine hoeveelheid vloeistof, waarvan de koncentratie geheel onbekend is, zodanig te wijzigen, dat deze vloeistof in één buisje met meerdere koncentraties van de vergelijkings-vloeistof wordt vergeleken.

## VII

De theorie van HEIJN betreffende de lengtegroei van de celwand verdient de voorkeur boven die van SÖDING.

Jahrb. wiss. Bot. 79, 1934, S. 753, 231.

## VIII

Het rusten der antennae van *Notonecta glauca* op de onder water meegevoerde lucht houdt geen verband met het percipieeren van evenwichtsstoringen.

## IX.

De aanwezigheid van *Flagellaten* in de darm van *Termieten* is geen symbioseverschijnsel.

MANSOUR, Biol. Rev., 9, 1934, p. 363

## X

Tot een bloem van *Zostera* kunnen worden gerekend een vruchtblad met de daaronder staande meeldraad.

MARKGRAF, Ber. d. d. bot. Ges., 14, 1936

## XI

De nectariën van *Salix* en het bekertje van *Populus* zijn op te vatten als een rudimentair perigoon.

## XII

Ten onrechte meent MEZ, dat de *Lentibulariaceae* bij de *Primulales* moeten worden ondergebracht.

Bot. Arch., 38, 1936, p. 86

## XIII

Het is beter de *Triglochineae* tot de *Potamogetonaceae* te rekenen, dan ze met *Scheuchzeria* in de *Juncaginaceae* onder te brengen.

Ber. d. d. bot. Ges., 14, 1936

CONTRIBUTIONS TO THE GENETICS OF
TENEBRIO MOLITOR L.

# CONTRIBUTIONS TO THE GENETICS OF TENEBRIO MOLITOR L.

PROEFSCHRIFT TER VERKRIJGING VAN DEN GRAAD VAN DOCTOR IN DE WIS- EN NATUURKUNDE AAN DE RIJKS-UNIVERSITEIT TE GRONINGEN OP GEZAG VAN DEN RECTOR MAGNIFICUS Dr. G. F. ROCHAT, HOOGLEERAAR IN DE FACULTEIT DER GENEESKUNDE, TEGEN DE BEDENKINGEN VAN DE FACULTEIT DER WIS- EN NATUURKUNDE TE VERDEDIGEN OP WOENSDAG 10 FEBRUARI 1937 DES NAMIDDAGS TE HALF VIJF PRECIES

DOOR

## JACOBES JOHANNES SCHUURMAN
GEBOREN TE TOLBERT

'S-GRAVENHAGE
SPRINGER-SCIENCE+BUSINESS MEDIA, B.V.
1937

ISBN 978-94-017-6715-6        ISBN 978-94-017-6794-1 (eBook)
DOI 10.1007/978-94-017-6794-1

PROMOTOR:
PROF. DR. T. TAMMES

*AAN MIJN OUDERS*

# CONTRIBUTIONS TO THE GENETICS OF TENEBRIO MOLITOR L.

by

J. J. SCHUURMAN

(Received for publication December 9th 1936)

TABLE OF CONTENTS

### INTRODUCTION

In 1915 ARENDSEN HEIN started experiments with *Tenebrio molitor* on a large scale. A large number of varieties were examined by him for their heredity, and he also determined what conditions and what food are most favourable for the culture of these animals. The comprehensive results were published in a series of papers during the years 1920–1924 (2, 3, 4, 5, 6).

After ARENDSEN HEIN's death the investigation was continued in Groningen by FERWERDA (29). I will summarize the results of their work here as far as they are necessary for the discussion of my own work.

ARENDSEN HEIN already ascertained that the three colour types: orange, yellow brown and umber brown, indicated according to the colour of the larvae, differ from each other in one gene. The sequence of dominance is orange → umber brown → yellow brown.

As FERWERDA communicates, a reversal of dominance occurs in the cross umber brown with orange. The larva looks more like the orange, the beetle like the umber brown type. FERWERDA fully discusses the origin of pigments and this has induced me to investigate what the difference between orange and umber brown consists in. (Chapter I).

The type with V-shaped head groove found by ARENDSEN HEIN and examined genetically by FERWERDA is dominant over normal and is controlled by one gene (B), which lies in the same chromosome as the factor g for flesh-coloured eyes. According to FERWERDA the factors B and g have a lethal effect, when present in a homozygous condition. This linkage and this lethality have been further examined by me (Chapter V).

The eye colour had already been partly analysed by ARENDSEN HEIN (3). FERWERDA found the following genetic formulae for the four varieties black, red, yellow and flesh-coloured:

| ♀ | ♂ | |
|---|---|---|
| FFGGHH | FFGGHo | black |
| ffGGHH | ffGGHo | red |
| FFGGhh | FFGGho | } yellow |
| ffGGhh | ffGGho | |
| FFggHH | FFggHo | } flesh-coloured |
| ffggHH | ffggHo | |
| FFgghh | FFggho | |

FERWERDA evidently did not know the triple recessive variety. I have tried to ascertain the eye colour of this genotype (Chapter II).

Besides these 4 types there is a fifth which FERWERDA called "gefleckt". The eye of these animals is partly black, partly red. In chapter III the results of a number of crosses are discussed, which I have made with a view to the analysis of this character.

One of my crosses gave rise to an animal with eyes which were partly black, partly flesh-coloured. This case has been analysed in chapter IV.

The great difference in length among individuals in the same culture induced me to trace whether these are due to genetic differences (Chapter VI).

Larvae, pupae and beetles were exposed to ultra-violet rays, in order to discover whether this treatment caused any mutations (Chapter VI).

As the technique has undergone but slight alterations, a short description may suffice.

The females deposit their eggs on bits of flannel specially provided for this purpose. These eggs are counted twice a week and then removed with the bit of flannel to little ointment pots standing in an incubator at 26° C. After about 10 days the young larvae emerge. These are counted once a week, and are then transferred to a big ointment pot with food, also kept in the incubator. After 5 to 7 months they pupate and they are then transferred twice a week to flat earthen plates. After about 10 days the beetle emerges, it is examined and transferred to a flat dish similar to the one in which the pupae are. If necessary males and females are separated. These dishes of pupae and beetles are placed on top of each other on shelves along the wall of the room.

For crossing purposes a male and a female or a number of males and females are put in a glass crystallization dish, which is then put in an incubator having a temperature of about 26° C. For food they get a piece of dry rusk, a slice of raw potato for moisture, and a little bit of paste consisting of equal parts of white of egg and rusk. All this is renewed twice a week. The larvae get the same food as mentioned by ARENDSEN HEIN (2); the beetles in the beetle-dishes get dry rusk. Both get a slice of potato for moisture, the larvae twice a week and the beetles once. In the pupae-dishes a slice of potato is also given once a week in order to prevent the young beetles from feeding on the pupae.

The research was made in the Genetic Institute of the State University at Groningen under the guidance of Prof. Dr. T. TAMMES.

# CHAPTER I

## PIGMENT FORMATION IN DARK BROWN AND MELANISTIC BEETLES

### § 1. *Introduction*

ARENDSEN HEIN (4, 5) already knew three colour types in *Tenebrio*. Afterwards these types have been analysed genetically by FERWERDA (29, p. 31), when the differences proved to be based on a series of multiple allelomorphs. He indicated this by using the following symbols:

| | |
|---|---|
| orange larva | dark brown beetle AA |
| yellow brown larva | dark brown beetle $a_1a_1$ |
| umber coloured larva | black beetle $a_2a_2$ |

assuming that these three factors control both the formation of colouring matter in the larva and in the beetle.

HAECKER (40) was the first to see that the material relation between gene and realized character should be found. GOLDSCHMIDT (36) has gone further in this direction and has, now that the factor analysis in various experimental objects is more or less advanced, insisted on our paying more attention to the developmental physiology of the characters we observe, in order to be able to penetrate more deeply into the substance of the genes.

DANNEEL (15) has communicated the results of the investigations concerning pigmentation of rabbits. He, however, could not state the chemical differences between these varieties.

As far as I know KÜHN and coworkers (49–50) have been the first to attain satisfactory results with such investigations on *Ephestia Kühniella*. They know an allelomorphic series A, $a^k$ and a, pleiotropically controlling the pigment formation in the hypodermis and stemmata of larvae and eyes, testes and „brains" of imagines. AA individuals have black eyes, brown-violet testes and brownish lobi-optici, their caterpillars possess a reddish hypodermis and strongly pigmented stemmata. An aa imago has red eyes, usually colourless testes and bright red lobi-optici. The hypodermis of caterpillars is colourless, the stemmata are faintly pigmented. Finally $a^k a^k$ individuals have coffee-brown eyes and bright coloured testes. The hypodermis of the caterpillar is colourless, the stemmata are faintly pigmented. A is dominant over $a^k$ and a.

If testes, ovaries or „brains" of an AA individual are transplanted to an aa individual after some time the pigmentation of the latter much resembles that of an AA individual. Thus these organs secrete a substance which influences pigmentation and is therefore an intermediary in the reaction series from gene to character. KÜHN has decided that this substance has the character of a hormone. The gene A causes the formation of a distinct hormone, the gene a the formation of an other and this is the reason of a difference in colour.

Of late WIT (86) has investigated the chemical differences between distinct colour types of Aster. He has stated that these colours arise from anthocyanidines called delphinidine, cyanidine and pelargonidine.

To be sure in *Tenebrio molitor* genetic analysis is only in its initial stage, but yet there is a possibility in this case to commence such an investigation. As already mentioned above the genetic differences between the colour types have been fixed and it is now important to discover what material differences exist between these three types. For this purpose, however, it is necessary to have a general insight into the process of pigment formation.

On this subject a number of papers have been published t.w. FÜRTH und SCHNEIDER (33), PRZIBRAM and collaborators (64–66), HASEBROEK (42), VERNE (79) and last not least SCHMALFUSS and collaborators (68–72). The result of all these investigations has been that the general insight has been arrived at pigment formation being based on the oxidation of colourless chromogens under the influence of ferments which are found in the haemolymph. HASEBROEK reports on this as follows: „Ich teile zunächst aus den BLOCHSchen Arbeiten die wesentlichen Grundlagen mit. Die Anfänge gehen schon auf SCHÖNBEIN zurück, der zuerst nachwies, dass die spontane Färbung zahlreicher Pflanzensäfte auf Oxydation beruht, welche durch fermentartige Körper — die Oxydasen und Peroxydasen — aus farblosen Chromogenen gebildet werden. Bis jetzt sind nach BERTRAND zwischen solchen pflanzlichen Oxydasen zu unterscheiden die Phenolase, welche alle möglichen Phenole und die entsprechenden Amine zu oxydieren vermag, und die spezifische Tyrosinase die ausser Tyrosin und seinen peptidartigen Verbindungen nur Parakresol angreift. Schon bald nach der Entdeckung der Beziehung der pflanzlichen Tyrosinase zur Pigmentbildung suchte man nach analogen Verhältnissen bei Tieren. Und man fand in der Tat gerade bei den niederen Tierklassen die Tyrosinase, zunächst im Darm des Mehlkäfers (BIEDERMANN) dann in der Hämolymphe von Schwärmerpuppen (V. FÜRTH und SCHNEIDER). Die Tyrosinase konnte aus dem Blut der Puppen von *Chaer. elpenor* und *Deil. euphorbiae* isoliert und in ihrer Reaktionsweise auf Tyrosin in vitro verfolgt werden, indem durch den Zusatz die wässrige Lösung von Tyrosin sich erst violett, dann schwarz färbte, worauf schliesslich es zur Ausscheidung von dunklen Flocken kam. Auch ist in neuerer Zeit das Tyrosin selbst im Puppenhaut von *Pieris brassicae* als mutmassliche Vorstufe des schwarzen Pigmentes der Puppenhülle festgestellt.

BLOCH zeigte, dass bei Menschen und höheren Tieren keine Tyrosinase vorhanden ist, sondern eine andere Oxydase, für die in analoger Weise als Muttersubstanz das 3,4-Dioxyphenylalanin angenommen werden muss, eine Substanz, die chemisch dem Tyrosin sehr nahe steht (= Dopa).

Behandelt man — nach BLOCH — überlebende Schnitte der Haut von Menschen und Tieren mit einer $1-2^0/_{00}$ wässrigen Lösung von Dopa, so treten an bestimmten Stellen dunkelbraune bis tiefschwarze Färbungen auf. Diese Dopa-reaktion beruht also darauf, dass das Dopa durch Oxydation und Kondensation sich in einen schwarzgefärbten Körper — das Dopa-melanin — verwandelt. Hervorgerufen wird diese durch die Dopa-oxydase, die ihren Sitz in den Elementen der Haut hat."

The ferment chromogen hypothesis may also be adopted for *Tenebrio*, as has appeared from SCHMALFUSS' experiments on this animal.

Among the chromogens may be classed in general all substances which pass by oxidation into a coloured matter, the presence of a certain ferment usually bringing about a rapid reaction. HASEBROEK found a method by which one can get such ferments at one's disposal outside the body of the animal in question. SCHMALFUSS worked out this method and gives a full account of it. He applies it as follows: with the aid of a small glass capillary with finely drawn out point haemolymph is withdrawn from the animal. For this purpose the point of the capillary is stuck between two chitin rings of the segments into the dorsal vessel. Through the capillary action the tube is filled with a more or less clear fluid. Next the point of the capillary is drawn across filter paper, so that a strip of a width of about 3 mms is soaked with this fluid. Then the filter paper is dried as quickly as possible in vacuo over phosphorouspentoxide, in order to avoid spontaneous melanosis and finally cut right across the strip soaked in ferment into strips of a width of about 1 mm, which I shall call experimental strips. If such an experimental strip is put in a chromogen solution, colouring matter arises exactly there where the ferment is found. With such an experimental strip SCHMALFUSS (66) was still able to show formation of colouring matter in an m/50,000,000 solution of 1, β 3, 4 dioxyphenol α amidopropionic acid. When they are put up air tight, dry and cool, such papers will keep good for at least 15 months.

The chromogen of the animal can also be obtained. SCHMALFUSS discovered that in the exoskeleton of various species of beetles, even after the animal has been fully darkened, there still remains an amount of chromogen which is not oxidized, because the circumstances are no longer sufficiently favourable. In young, just emerged and therefore still nearly uncoloured beetles the existence of this chromogen could not be proved. The quantity of chromogen increases as the exoskeleton of the animal grows harder and darker in the course of its development.

In order to obtain this chromogen the exoskeleton, or when smaller beetles were concerned, a number of exoskeletons or elytra were pounded in a mortar and next mixed with water, put in a boiling water-bath for about 2 minutes. On filtration a clear fluid appeared to which an experimental strip with ferment was added, which in the presence of a sufficient amount of chromogen assumed colour. The colour intensity of the experimental strips was a standard for the quantity of chromogen present.

If we assume that neither the ferment nor the chromogen have been altered, it is possible to imitate the process of pigment formation as it is enacted in the animal with the aid of a chromogen solution and an experimental strip. Now the possibility also exsists to investigate the difference between the orange dark brown type and the umber coloured melanistic type.

For briefness' sake I shall call these varieties in future respectively dark brown and melanistic. Here the question therefore rises on what this difference may be based. FERWERDA (p. 55) gave four possibilities; they may have: 1. the same ferment, but different chromogens; 2. different ferments and the same chromogen;.3. different ferments and different chromogens; 4. the same ferment and the same chromogen. He deems the one mentioned sub 1 the most likely on the strength of the fact that HASEBROEK (41) found in *Cymatophora or* that the ferment of the melanistic type and that of the non-melanistic one affect tyrosine and dopa in the same way, whereas the experimental strips with ferment are coloured more strongly by the integument of the melanistic than by that of the non-melanistic type.

The possibilities mentioned sub 2 and 3 FERWERDA leaves out of account, because no instances of them are known in literature.

According to him the following facts tell against the fourth possibility. If the pigment of the melanistic and of the dark brown beetle arises from the same chromogen under the influence of the same ferment, while in the melanistic type the chromogen only as a result of some attendant circumstance is oxidized more strongly than in the dark brown beetle, it is self-evident that the melanistic type during the first stages of darkening will subsequently show the same intermediate shades as the dark brown beetle. This, however, is by no means the case. Besides the melanistic type is darkened more rapidly than the dark brown beetle.

Especially the latter argument seems to me to count.

By making a number of experiments I have tried to ascertain which of the four possibilities is the correct one.

I shall now proceed to discuss these experiments, beginning with those which have not led to a positive result.

### § 2. *Experiments with paper strips soaked in ferment*

By means of different combinations of ferments and chromogens, we may perhaps discover what the difference consists in. The following combinations are possible: 1. ferment of the dark brown beetle with chromogen of the dark brown beetle; 2. ferment of the melan-istic beetle with chromogen of the dark brown beetle; 3. ferment of the dark brown beetle with chromogen of the melanistic beetle; 4. ferment of the melanistic beetle with chromogen of the melanistic beetle.

These experiments have been made in the following way:

Ferment and chromogen were obtained in the way indicated by SCHMALFUSS. If necessary the chromogen was divided into a number of equal portions, all of which were put in a separate dish, after which one experimental strip with ferment was added to each. A number of these small porcelain dishes were subsequently placed in a large glass dish, at the bottom of which there was a little water, in order to restrict the evaporation of the chromogen solutions. The whole was covered with a glass plate and put in an incubator at a temperature of 26° C.

After a specified time the experimental strips were washed out in distilled water, dried between filter paper and next the colour was compared.

In the first experiment all possible combinations were made be-tween chromogens of dark brown and of melanistic beetles and experimental strips with ferment of dark brown and melanistic beetles and beetles of *Tenebrio syriacus*. After $41\frac{1}{2}$ hours the 6 ex-perimental strips were dried and compared. The colour of all of them was chocolate brown, corresponding with No. 110 of the Code des Couleurs of KLINCKSIECK et VALETTE (48), also those of the combi-nation of melanistic chromogen and melanistic ferment.

Considering the possibility that this was due to lack of chromogen, I at once added a fresh melanistic experimental strip to the same melanistic chromogen, after the old experimental strip had been re-moved. After 96 hours a distinct pigment formation was to be ob-served, which points to the fact that a fair quantity of chromogen was left and deficiency of chromogen can therefore not have been the cause of the non-occurrence of melanistic colouring matter.

In spite of this result another experiment was made with a melan-istic chromogen solution from 4 times as many elytra. After at least 48 hours the experimental strip added to it, had assumed the same colour as the corresponding one in the previous experiment.

In order to avoid that the melanistic chromogen might possibly be altered by the boiling water-bath, an experiment was made with chromogen which was extracted in the cold from 89 mgrs elytra during 48 hours. This yielded a light brown pigment both with an experimental strip with ferment of a dark brown and of a melanistic beetle.

Besides on the above chromogens, the action of the ferments has been tested on pyrocatechin, tyrosine, phenol, hydroquinone, dopa and on chromogen of beetles of *Tenebrio syriacus*. Result: no difference between the action of the two ferments. In the condition in which they are after absorption by filter paper therefore, these ferments exercise the same influence on all the above mentioned substances.

Neither has a difference been shown between the chromogens.

However, it has already been indicated above for what reasons it cannot be assumed that both the ferments and the chromogens of the two varieties are identical. These negative results led to our instituting a new series of experiments, which is more in correspondence with the process in the body of the beetle.

§ 3. *Experiments with ferment in solution*

It is conceivable that the ferment outside the body of the beetle and especially in the dried condition in which it is in the filter paper, has changed. To avoid this, directly on being taken from the animal, the ferment being dissolved in the haemolymph in this series of experiments was added to the chromogen. The chromogen was extracted in the same way as mentioned above. For the rest the experiment was arranged as in the case of the experimental strips.

*a.* Negative results.

An extensive experiment was made with chromogen solutions from 304 mgrs of melanistic elytra and of as many dark brown ones, both in 5 ccs distilled water. The two quantities were divided into 5 equally large portions, after which to the chromogen of the melanistic type ferment was added of respectively an orange larva, a dark brown beetle, an umber brown larva and a larva and a beetle of *T. syriacus*. To the chromogen of the dark brown beetle ferment was added of respectively an orange larva, an umber brown larva, a melanistic beetle and of a larva and a beetle of *T. syriacus*. The result was perfectly equal in all the 10 cases t.w. a dark brown colouring matter.

The same result was obtained with two solutions of respectively 188 mgrs of melanistic and dark brown elytra, both of which were divided into two portions. To the portions there was respectively added dark brown and melanistic ferment. After evaporation all 4 precipitates were of a colour about equal to that of the dark brown beetle.

Hitherto the pH of the chromogen solutions had not been taken into account. According to SCHMALFUSS pigment formation is optimal at a pH of 7.8–8.3. In the then following experiment 2 solutions were made each of 85 mgrs of melanistic elytra and 2 each of 85 mgrs of dark brown elytra. These solutions proved to have an acid reaction and were therefore neutralized by much diluted NaOH, to which phenolred was added as an indicator. This was done by adding the Na OH which was coloured red by the phenolred drop by drop to such an amount that the phenolred was no more decolorized. Next ferment was added, so that the same combinations were made as in the experiment described above. The result was nevertheless identical with that of the above experiment.

*b.* Positive results.

The idea that after all boiling would have a certain influence on the chemical composition, at any rate on the „Reaktionsbereit-schaft" of the chromogens, however, left me no peace. SCHMALFUSS states that after about 2 minutes a large portion of the chromogen has been dissolved. He does not state, however, whether longer extraction has a detrimental effect, on the contrary he gives the impression that this is not the case, by adding that through extraction with hydrochloric acid the rest of the chromogen can also be obtained. Moreover he writes (1929 p. 81): „Ist aber die Konzentration am M sehr gering, so wird man möglichst viel Pulver mit heissem Wasser behandeln und das Filtrat eindampfen müssen."

So far I had not paid much attention to the duration of the extraction. Therefore another experiment was made, in which extraction was continued for exactly 2 minutes. It was carried out with two chromogen solutions of respectively 85 mgrs of dark brown and 87 mgrs of melanistic beetle elytra, to which ferment of respectively dark brown and melanistic beetles was added. After 24 hours the combination of melanistic chromogen and ferment was coloured distinctly darker than that of dark brown chromogen and ferment. This difference was afterwards maintained and it could be clearly stated that it was not based upon a difference in amount of pigment. Now a second experiment was started in order of solving the question of the difference between dark brown and melanistic. Two extracts

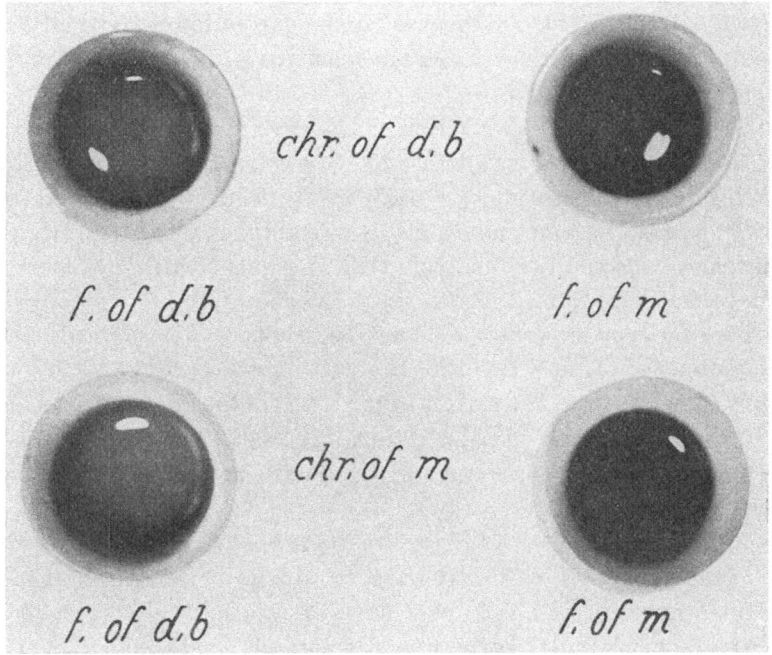

FIG. 1. Reaktion of ferments of dark brown and melanistic beetles on chromogens of dark brown and melanistic beetles.

    chr = chromogen                          f = ferment
    db  = dark brown                          m = melanistic.

were made, each of 30 dark brown elytra and two each of 30 melanistic ones. To the one of the dark brown beetle dark brown ferment was added, to the other ferment of the melanistic beetle. In the same way the two melanistic extracts were treated. After 48 hours a photograph was taken of this experiment (see fig. 1). It is clearly visible that the ferment of the melanistic one affects the chromogen of both the melanistic and the dark brown beetle in the same way.

The action of the ferment of the dark brown beetle on the two chromogens is likewise the same, but the action of the two ferments is quite different, t.w. the colour caused by the ferment of the melanistic one is much darker than the one due to the ferment of the dark brown beetle. From this it follows that there is a difference between the ferments of the dark brown and of the melanistic variety.

Besides it follows from this experiment that we should be careful when extracting the chromogens. If the extraction is continued for more than 2 minutes we risk the occurrence of alterations in the chromogens.

The action of the two ferments, however, was also ascertained with respect to pyrocatechin. To one of two equally large quantities of pyrocatechin ferment of a dark brown beetle was added, to the other ferment of a melanistic one. Between the two there was a clear difference, also on our repeating the experiment. After evaporation the precipitate formed under the influence of the ferment of the melanistic type appeared to be nearly black; the one formed under the influence of the ferment of the dark brown beetle, dark brown. This is therefore a confirmation of the result of the above experiment.

So it can now be considered an established fact that there is a difference between the ferments of the dark brown beetle and that of the melanistic variety.

As to the chromogens I have not been able to show a difference between them. The fact that owing to different ferments different pigments arise from one and the same chromogen (pyrocatechin) as in the above mentioned experiment, points to the possibility that the chromogens may be the same. It is, however, not a proof of their being identical.

As an objection to the equality of the chromogens might be alleged the lack of common stages of darkening, which in this case might be expected. However, the structure of a chromogen molecule is so complicated, that it is altogether possible that an absorption of oxygen can take place at one of several points. Under the influence of one ferment the oxygen might be absorbed at one definite point and under the influence of an other ferment moreover or solely at another point with the result that two substances are formed which are coloured differently. In that case common stages of darkening could not be expected. Besides the fact that no common stages of darkening are to be seen does not prove that they do not exist. If we assume that under the influence of the ferment of the melanistic beetle oxygen is taken up more rapidly and may be after all in a greater quantity, it is evident that of the common stages of darkening but little will be visible, because the chromogen first supplied in the melanistic beetle will not only be rapidly converted into black pigment, but this first formed black pigment will render it impossible for us to observe in all its stages the conversion into pigment of chromogen supplied later.

By the above I hope I have succeeded in showing that the existence of identical chromogens is possible. In my opinion the theory of identical chromogens and different ferments is even to be preferred to that of different ferments and different chromogens, also considered from a genetic point of view. Genetically the difference between melanistic and dark brown consists in one gene. It seems improbable to me that the material difference as a result of the genetical one, would consist in a difference both in ferment and chromogen. From this we cannot but infer that one of the series of allelomorphs would control the formation of a certain ferment and a certain chromogen, on the other hand an other allelomorph the formation of an other ferment as well as of an other chromogen. Much simpler is the assumption that as a result of the monofactorial difference, there has only appeared a difference between the ferments.

In addition it is a remarkable fact that such a great correspondence exists between the chromogens and pyrocatechin. Above it has been shown that under the influence of the ferments of dark brown and melanistic beetles two different pigments are formed from pyrocatechin. The correspondence, however, is still greater. When a solution of pyrocatechin is kept in a boiling water-bath for some minutes, its reaction to both ferments has become identical. Both under the influence of ferment from a dark brown beetle and of ferment from a melanistic beetle a dark brown precipitate is formed.

### § 4. *Experiments with injections*

During the above mentioned experiments a series of other experiments was made, the result of which supports the assumption that the chromogens of the melanistic and of the dark brown types are equal.

Starting from the fact that the most ideal condition for the formation of pigment is found in the body of a young beetle, a number of these young, not yet darkened beetles of the dark brown variety were injected with chromogen.

For the injections a glass tube with a finely drawn out point was used. Over the other end of the tube a rubbertube was slipped, closed at the top. By means of this a liquid could be sucked up in the tube and pressed out of it. The beetles were pricked in the lower end of the thorax or in the abdomen and subsequently injected; pupae were injected in the abdomen.

Altogether 36 beetles were injected with melanistic chromogen in the thorax, 23 of which showed a black spot after 24 hours. In 10 beetles which were injected in the abdomen no stain appeared. 9 beetles were injected with dark brown chromogen; 4 of them showed a stain. All the pupae died.

In order to check the results 10 beetles were injected with water from the water supply and 8 only pricked. In those no stain was observed, so that we may assume that the dark stains which arose after injection with chromogen were indeed due to oxidation of that chromogen.

Now the question rises whether there is a difference between the stains caused by the chromogen of dark brown beetles and that of melanistic beetles. On examination with a magnifying glass it appeared that both consisted of a large number of dark dots, but a difference was not to be seen. In this case too therefore we did not succeed in demonstrating a difference between the two chromogens.

Injecting young melanistic beetles has been omitted, because they are darkened much quicker than dark brown ones. After 24 hours a stain that might have arisen would not be perceptible.

Nor have injections with ferment been given. According to SCHMALFUSS chromogen is the inhibitory factor during darkening, for must not it be supplied again and again, whereas ferment is constantly present? When additional chromogen is locally supplied, the process of darkening can take place more rapidly on that spot with the result that a dark spot will arise there. The case is quite different, however, on addition of extra ferment. The process is not accelerated by it in any respect and a dark stain is not to be expected, whence these experiments have not been made.

## § 5. *Difference between larva and beetle*

Here too there are theoretically 4 possibilities, viz. 1. larva and beetle are the same, 2. they only differ in ferment, 3. they only differ in chromogen, 4. they differ in both.

HASEBROEK (42) found in *Cymatophora or* the same oxidases in the haemolymph of egg, caterpillar, pupa and butterfly.

This seems to me to be most likely in *Tenebrio*. For as there are no genetical differences between larva and beetle, we should have to assume that the same gene in the larval stage would influence the production of a certain ferment and then give suddenly rise to a different ferment in the beetle stage. The same also obtains mutatis mutandis for the possibility mentioned sub 3 and in a still stronger degree for that mentioned sub 4.

Should, however, a difference in ferments or chromogens exist after all, the most likely explanation seems to be that there is a factor which can not express itself before the beetle stage.

If indeed the beetle and the larva possess the same chromogen and the same ferment, it may be expected that the beetle, since it is darker than the larva, will especially in the very beginning develop colours which are identical with colour occurring in the larva. The different colours of the larva (see FERWERDA pl. I, figs. 1, 2, 3) are to be found back in the different stages of darkening in the beetle.

On comparing young beetles, both of the dark brown and of the melanistic type in various stages of darkening to larvae, we find that in the former type the colour of the larva is slightly less red than that of the beetle, viz. somewhat more grayish. When on the other hand a melanistic larva and a young semi-darkened melanistic beetle are compared, it appears in a certain stage that the colour of the dark rings of the larva is perfectly equal to the colour of certain parts of the beetle body. The lighter shades of larva and beetle also correspond.

In my opinion it may be assumed that when in the dark brown variety ferment and chromogen of the larva are identical with those of the beetle, this also may be the case in the melanistic variety. If there is a difference this is likely to be shown in both varieties. For this reason and also because there was a greater supply of material of the dark brown variety, the experiments which have been made with reference to this matter, were exclusively made with the dark brown variety.

These experiments will not be discussed so fully as the preceding; let it be sufficient to state the results. The ferment of the larva as well as that of the beetle was combined in the shape of experimental strips with chromogen of orange larvae, dark brown and melanistic beetle and of the beetle of *T. syriacus*, pyrocatechin, phenol, tyrosine, and hydroquinone without any difference showing between two corresponding experimental strips.

Also the chromogens of orange larva and darkbrown beetle reacted in the same way.

Experiments have also been made with ferments in solution in the way discussed above. On our adding pyrocatechin both yielded pigment of a dark brown colour. As in the corresponding experiment discussed above, differences were shown between the ferments of dark brown and melanistic, I think I may assume that there is no difference between the ferments of larva and beetle.

# CHAPTER II

## THE TRIPLE RECESSIVE FORM OF EYE COLOUR

FERWERDA states that the eye of the genotypes FFgghh and ffggHH is flesh-coloured. As he did not know the genotype ffgghh, I have tried to obtain it in order to be able to determine its eye colour.

I started with crossing a yellow-eyed ♀ and a ♂ with flesh-coloured eyes. The $F_1$ consisted of 49 beetles, all recorded as yellow-eyed. Probably, however, half of them were yellow-eyed and half of them flesh-coloured, for on being examined once more, when, however not all beetles were left, there were also found individuals with flesh-coloured eyes, which on first being examined were entered as yellow-eyed, in consequence of the great resemblance. From the fact that in this $F_1$ no black-eyed individuals arose, it may already be inferred that the flesh-coloured ♂ did not possess the factor H. The formulae of the parents evidently were FFGghh and FFggho.

As this conclusion was only drawn later on, it has not been taken into account on mating the $F_1$ animals. Two crosses were made, viz. a mixed cross of yellow-eyed and flesh-coloured individuals (Table 1, 355 A) and an intercrossing of flesh-coloured individuals (Table 1, 355 B).

TABLE 1. CROSSES WITH $F_1$ INDIVIDUALS

| number | yellow | | flesh-col. | | total | |
|---|---|---|---|---|---|---|
| | ♀♀ | ♂♂ | ♀♀ | ♂♂ | ♀♀ | ♂♂ |
| 355 A | 34 | 22 | 31 | 35 | 65 | 57 |
| 355 B | | | 19 | 19 | 19 | 19 |

By the cross yellow-eyed × flesh-coloured from 355 A, it was once more confirmed that the flesh-coloured individuals lack the factor H (Table 2).

TABLE 2. PROGENY OF YELLOW × FLESH-COLOURED

| number | yellow | | flesh-col. | | total | |
|---|---|---|---|---|---|---|
| | ♀♀ | ♂♂ | ♀♀ | ♂♂ | ♀♀ | ♂♂ |
| 355 C | 37 | 28 | 14 | 15 | 51 | 43 |
| D | 38 | 31 | 18 | 14 | 56 | 45 |
| E | 15 | 24 | 7 | 7 | 22 | 31 |
| total | 90 | 83 | 39 | 36 | 129 | 119 |

So the beetles obtained with flesh-coloured eyes are genotypically FFgghh.

In order to eliminate the factor F, flesh-coloured ♂♂ were crossed with red-eyed ♀♀. This cross yielded 35 ♂♂ + 37 ♀♀, all of them black-eyed, from which it follows that the factor F was indeed present. All of these black-eyed individuals therefore had the formula FfGgHh or FfGgHo and were mated inter-se (table 3). In this case a ratio was expected of 27 black : 9 red : 12 yellow : 15 flesh-coloured : 1 triple recessive.

TABLE 3. PROGENY OF BLACK-EYED BEETLES

| number | black | | red | | yellow | | flesh-col. | | total | |
|---|---|---|---|---|---|---|---|---|---|---|
| | ♀♀ | ♂♂ | ♀♀ | ♂♂ | ♀♀ | ♂♂ | ♀♀ | ♂♂ | ♀♀ | ♂♂ |
| 355 G | 32 | 26 | 9 | 3 | | 19 | 16 | 12 | 57 | 60 |
| | 58 | | 12 | | 19 | | 28 | | triple rec. | |
| theor. | 49.35 | | 16.45 | | 21.94 | | 27.42 | | 1.83 | |
| m | 5.34 | | 3.76 | | 4.23 | | 4.58 | | | |
| D/m | 1.56 | | 1.18 | | 0.67 | | 0.13 | | | |

A new eye colour has therefore not been found. Theoretically one ♂ of the formula ffggho could be expected as a result of the fact that the factor h is sex-linked. As the chance of actually obtaining this ♂ is exceedingly slight, the red-eyed individuals obtained were mated inter-se and subsequently their red-eyed offspring. As these two crosses are the same they will be discussed together (Table 4).

TABLE 4. PROGENY OF RED-EYED INDIVIDUALS

| number | black | | red | | flesh-col. | | total | |
|---|---|---|---|---|---|---|---|---|
| | ♀♀ | ♂♂ | ♀♀ | ♂♂ | ♀♀ | ♂♂ | ♀♀ | ♂♂ |
| 355 H | 2 | 3 | 50 | 60 | 5 | 17 | 57 | 80 |
| 355 I | 1 | | 210 | 163 | 9 | 24 | 220 | 187 |
| total | 3 | 3 | 260 | 223 | 14 | 41 | 277 | 267 |

The theoretical ratio cannot be given, as the genotype of the red-eyed individuals used was not the same in all of them. Yellow-eyed individuals could surely be expected. Probably they have been entered as flesh-coloured for the reason further explained in chapter III. In this cross there may theoretically arise only yellow-eyed ♂♂ and no yellow-eyed ♀♀ and as many flesh-coloured ♂♂ as ♀♀. In both progenies, however, we note a great excess of flesh-coloured ♂♂ over ♀♀. I think I may assume that this excess was in reality yellow-eyed.

But what is the real point here is, that no new eye colour has been found, though theoretically 7 individuals could be expected of the formula ffggho, viz. half of all ♂♂ of the formula gg, as a result of the fact that the factor h lies in the sex-chromosome. In this computation the number of ♂♂ is considered to be equal to that of ♀♀. So I think I may assume that the genotype ffgghh is flesh-coloured. For an exact proof special crosses ought to be made with a number of flesh-coloured ♂♂. For lack of time this had, however, to be omitted.

As contrasted with what MORGAN and BRIDGES (54) found in *Drosophila*, and ANNA R. WHITING (84) in *Habrobracon*, the triple recessive type is not coloured lighter than the simple or double recessive types.

With reference to these crosses I wish to make some physiological remarks. From SCHMALFUSS' researches (68–72) it appears that there is reason to assume that the pigments in the eyes arise in the same way as those in the exoskeleton, i.e. through oxidation of a chromogen under the influence of a ferment. We now know that in *Tenebrio* the following genotypes have the eye colour:

| | | |
|---|---|---|
| FFGGHH | FFGGHo | black |
| ffGGHH | ffGGHo | red |
| FFGGhh | FFGGho | } yellow |
| ffGGhh | ffGGho | |
| FFggHH | FFggHo | |
| ffggHH | ffggHo | } flesh-coloured |
| FFgghh | FFggho | |
| ffgghh | ffggho | |

In connection with these formulae, I arrive at the hypothesis: that the four last mentioned types have flesh-coloured eyes is a result of a lack of chromogen. Under the influence of the gene G a chromogen is formed which when it is not oxidized, gives a yellow colour to the eyes. This chromogen is oxidized under the influence of the gene H, which renders the eyes red and under the influence of the gene F, but only when the gene H is present, this chromogen is oxidized to a black pigment. If H is absent, F can do nothing.

In this hypothesis it is therefore assumed that under the influence of the genes F and H oxidative ferments are formed.

This hypothesis may account for: 1. the eye colour of the above mentioned 8 genotypes; 2. the turning red of the originally yellow eyes; 3. the turning black of the originally red eyes; 4. the remaining constant of the flesh-coloured eyes.

In the yellow eye there is chromogen, but no oxidizing ferment. Oxygen, however, can be absorbed and although more slowly, the yellow chromogen is yet oxidized and finally the eye grows red. The same may apply mutatis mutandis to the getting black of the red eye. That the flesh-coloured eye cannot change its colour is self-evident, for there is no chromogen present.

I, however, cannot account for the fact that the yellow eye of an FFGGhh individual turns red more quickly than that of an ffGGhh individual. Perhaps some insight may be obtained in this matter with the aid of HASEBROEK's and SCHMALFUSS' methods.

# CHAPTER III

## THE MOSAIC EYE

### § 1. *Literature*

In various animals hereditary mosaic phenomena have been observed (STUBBE 76, 1933). Often the occurrence of two colours was concerned, usually irregularly mixed; sometimes, however, the occurrence of a different shape or size of special parts of the body.

In *Drosophila* a large number of mosaics are known. A mosaic individual often spontaneously occurred among a number of normal individuals; many cases, however, were due to X-raying.

SPENCER (74, 1926) cultivated a strain of purple-white flies. In this strain individuals appeared with red ommatids in the otherwise white eye. Such individuals yielded mottled individuals when mated together, but varying greatly in number and in degree.

PLOUGH (63, 1927) had *Drosophilae* with a recessive, sex-linked gene, which suppresses black body colour rendering such flies phenotypically to be black. This gene, however, often mutates back to normal, which gives rise to mosaic individuals.

WEINSTEIN (quoted after MULLER 57, 1930) found an eye colour dominant over normal. This colour, however, was eversporting as a result of a translocation.

MULLER (58, 1930), in his experiments with X-rays, found a number of cases of eversporting characters which he called mottled. He thinks they are due to displacements of parts of chromosomes, i.e. some to inversion, others to translocation. Afterwards he found 5 more cases, which he called Plum, Discoloured, Moiré, Tarnished and Punch coloured respectively, all going together with inversion or translocation.

In an X-ray experiment in 1926 he had also found an individual with mosaic bristles. A couple of bristles were modified into forked, the others were normal. He does not give an explanation, but he does give some hypotheses, adding, however, that they are speculative.

After X-raying VAN ATTA (7, 1932) found 7 dominant eye colour mutations, two of which were homogeneous, viz. Salmon and Henna, while the 5 others were eversporting. In some ommatids the normal red was diluted as in cream eye, found by HANSON and WINKELMAN. These dilute eye colours probably form an allelomorphic series in the extreme right end of chromosome II. Parallel with it goes the presence of chromosomal rearrangements.

PATTERSON (61, 1932) found a mottled individual in the progeny of an X-rayed ♂. From crosses it appeared that this was due to an unstable translocation: the broken left end of the X-chromosome had attached itself to IV. Because during the somatogenesis the translocated part with the dominant gene sometimes disappears, the recessive gene can utter itself in some cells. PATTERSON assumes that this piece disappears, as it does not take part in a somatic division.

SHULTZ (73, 1932) found a case similar to PLOUGH's. The eye colour vermilion can be suppressed by a suppressor, which renders the eye wild type. The action of this suppressor can be inhibited by a duplication, which belongs to the group of eversporting chromosome rearrangements found by DOBZHANSKY. This gives rise to mosaic individuals.

GLASS (35, 1934) studied 6 allels of brown (bw, II, 104.3). Phenotypically they all resemble Plum, described by MULLER (58). The pigmentation shows darker and lighter spots. In a homozygous condition all the 6 are almost entirely lethal. Two (Plum and Discoloured) go together with inversion, the four others (Tarnished, Rosy, 143 a and A 34) with mutual translocations.

In *Drosophila virilis* DEMEREC (16, 1926) saw the body colour reddish suddenly occur. Reddish appeared to be an allel of the sex-linked gene yellow, which often mutates to wild type, but only during the maturation division of heterozygous females and not in somatic cells. Back-mutations do not occur. In a reddish strain wild type individuals can therefore arise.

Afterwards (17) he found something similar regarding the gene for miniature α concerning shape and size of the wings. Miniature α repeatedly mutates to wild type, the reverse does not take place. These mutations, which also occurred in homozygous miniature α individuals, appeared during the formation of gametes and also in somatic cells in consequence of which mosaic individuals also arose. These mutations are influenced by other genes which fact enabled him to isolate a low mutable, a mosaic and a high mutable line.

Finally he found (18, 1927) a third, frequently mutating character, viz. magenta α. This concerns the eye colour and corresponds with miniature α, because mutations to the wild type also occur in this type during the reduction division of homo- and heterozygous ♀♀ and ♂♂ and in a slight number in somatic cells. Homozygous magenta α ♀♀ are largely sterile.

In *Drosophila hydei* SPENCER (75, 1930) examined the behaviour of mosaic orange, consisting of orange spots in vermilion eyes. In one culture there arose 14 of these individuals at a time with great variations in shape and size of the orange coloured portion. In part of the progeny mosaic orange likewise occurred, both in those of somatic normal and of somatic mosaic individuals. Temperature and moisture did not affect this number.

He gives two hypotheses for explanation:

1. mosaic orange is a recessive autosomal gene, phenotypically indistinguishable from wild type, but showing in somatic cells an inclination to mutate to the condition in which the gene acts as a modifier of vermilion,

2. mosaic orange is an extremely inconstant and variable character.

GREB (38, 1933) investigated the influence of the temperature on the origin of mosaic individuals in *Habrobracon*. These arise from binuclear eggs. He found that a constant, low temperature inhibits the origin of mosaic individuals, whereas a high temperature increases their number.

WHITING (82, 1933) found that in *Habrobracon* combination of the genes sv (shot veins, wing character) and wh (white eyes) gives rise to eyes which show red dots on a white ground. Variegated in a less degree also arises in the combination of the gene sv with the eye colour genes O (black), o$^d$ (dahlia), o (orange) and wh$^c$ (carrot).

CASTLE (12, 1912) traced the course of heredity in tricoloured

*Caviae*. The tricoloured animal is white with irregular but distinct black and yellow patches. Tricoloured animals constantly yield a tricoloured progeny and in addition black-white and yellow-white animals, which in their turn may also produce tricoloured progeny.

IBSEN (45, 1916) discussing these experiments gives CASTLE's explanation: „Now the tricolor race is a yellow one spotted both with white and with black i.e. it results from irregularity in distribution through the coat of two different chemical substances, the color factor and the black factor. These two factors are known to be independent of each other in heredity (see Castle 1909). It is therefore not to be supposed that they will coincide in distribution. If the black factor falls only on areas which lack the color factor it will produce no visible effect, and the animal will be yellow and white. If finally the black factor falls on some of the colored areas but not all of them, those in which it falls will be black, the others yellow and the uncolored areas of course white. Hence a tricolor will result."

IBSEN (45, 1916) himself gives a factorial explanation of the tricolour in *Caviae* just as for the tricolour in the basset hound.

I will wind up with mentioning some cases in plants. DE HAAN (39, 1933) discussed a number of them more fully.

EMERSON (24, 1929) assumes that the variegated pericarp of maize is due to somatic mutation.

Likewise IMAI (46, 1930) assumes that the green spots on the yellow leaves of *Pharbitis nil* arise as a result of mutation in a late stage in the ontogeny of the leaves.

CLAUSEN (quoted after PATTERSON 61, 1932) described a carmine-coral variegation in tobacco and is of opinion that this is due to an occasional loss of a piece of chromosome as a result of non-disjunction.

DEMEREC (20, 1931) found two mutable genes in *Delphinium ajacis*, viz. rose alpha and lavender alpha, both concerning the flower colour. Rose alpha often mutates to the purple wild type allel just like lavender alpha. Yet there is a great difference between the two, as rose alpha has a constant degree of mutability during the development of the petals and about the same degree during the formation of the gametes. The lavender alpha gene on the contrary is highly mutable in the early stages of development of the plant, but slightly mutable during the early stages of development of the perianth, while it is becoming again highly mutable towards the end of the development. In this article DEMEREC opposes the genomeric theory of EYSTER.

Finally IMAI and KANNA (47, 1935) found a form with striped yellow and striped creamish flowers in *Portulaca grandiflora*. They found striped yellow to be heterozygous and striped creamish homozygous recessive. Striped yellow self-pollinated yields 1 orange : 2 striped yellow : 1 striped creamish; striped creamish yields 1.37% orange, 13.51% striped yellow and 85.12% striped creamish. The explanation is that the recessive gene striped creamish is labile, and often mutates to dominant.

## § 2. *Description and occurrence*

FERWERDA (29, pp. 83, 84) remarks on this: „Der vierte Typus ist der gefleckte Typus, Pl. I, 10. Im Auge sind hier schwarze und rote Teile zu unterscheiden. In den meisten Fällen überwiegt das Rot, kann man also von schwarzen Flecken in einem übrigens rotfarbigen Auge reden; zuweilen aber überwiegt auch das Schwarz. Die Grenze der schwarzen Flecken trifft nicht mit derjenigen der Fazetten zusammen, m.a.W. man kann hier nicht von einer Gruppe schwarz pigmentierter Ommatidien in einem übrigens roten Auge reden. Es ist mir immer aufgefallen, dass das Schwarz seine stärkste Entwicklung in der ventralen Augenhälfte hat. Zwischen diesen beiden Extremen finden sich alle erdenklichen Übergänge. Das schwarze Pigment tritt, wie ich feststellen konnte, schon ziemlich früh im Pupalstadium auf, gewiss nicht später als beim normalen schwarzen Auge". (S. 81) „Bei dem eben ausgeschlüpften Käfer heben sich diese noch schwarz gefärbten Teile stark gegen die noch sehr hellfarbigen roten Teile ab; in diesem Stadium ist der gefleckte Typus am stärksten ausgeprägt. Wenn der Käfer älter wird verschwindet der starke Gegensatz zwischen schwarz und rot, indem letztere Farbe immer dunkeler wird und sich zu dunkel rotbraun verfärbt. Bei einem 14 Tage alten Käfer, der beim Ausschlüpfen sehr typische gefleckte Augen hatte, konnte ich die schwarzen Teile kaum noch von der anfänglich rotfarbigen unterscheiden."

I have little to add to this. As for the extension of the black pigment in the eye, it has struck me that nearly always that part of

the eye continues to be red for the longest time, that lies posterior
to the chitin-plate that penetrates the eye anteriorly. This chitin-
plate divides the eye imperfectly into a larger ventral and a smaller
dorsal part. It is as if two centres existed, one at the top and one
at the bottom of the eye, from where the formation of the black pigment is started.

Indeed the mosaic-eyed type is already to be recognized in the
pupa-stage, but only during the last few days. Unlike FERWERDA I
do not believe that the black in the mosaic eye of a newly emerged
beetle is identical to that in the eye of a normal black-eyed beetle.
In the black colour of the mosaic eye of a very young beetle there is
always a red reflection. In the black-eyed beetle the black may vary
a little, but on the whole it is deep black.

The mosaic eye was first observed by ARENDSEN HEIN in a cross
made by him in 1920 of a red-eyed ♀ and a black-eyed ♂. The $F_1$
consisted of 4 black-eyed ♀♀ and 4 black-eyed ♂♂, which were bred
together and yielded an $F_2$ of 165 individuals. Of those 59 ♀♀ and
62 ♂♂ were black-eyed, 22 ♀♀ and 15 ♂♂ red-eyed and 7 ♀♀ mosaic-
eyed. Here therefore mosaic-eyed individuals occurred the first time.
From this $F_2$ 3 mosaic-eyed ♀♀ were crossed with 1 red-eyed ♂, which
cross yielded 27 ♀♀ and 25 ♂♂, all red-eyed and 19 ♀♀ plus 39 ♂♂, all
mosaic-eyed. The mosaic-eyed beetles were used to obtain a mosaic-
eyed stock. From the generation obtained from this, again mosaic-
eyed individuals were taken for breeding a following generation and
so on, up to and included an $F_{13}$. The results of this have been stated
in table 5 and will be discussed in the following paragraph.

### § 3. *Mosaic-eyed beetles bred together*

ARENDSEN HEIN already knew that from intercrossing mosaic-
eyed individuals in each case red-eyed individuals arise. FERWERDA
confirmed this later on. After that time the number of crosses among
mosaic-eyed beetles has been greatly extended, as shown in the subjoined table. ARENDSEN HEIN's observation was completely corroborated: a pure stock of mosaic-eyed individuals has not yet been
obtained thus far.

TABLE 5. OFFSPRING OF MOSAIC-EYED BEETLES

| | black ♀♀ | black ♂♂ | mosaic ♀♀ | mosaic ♂♂ | red ♀♀ | red ♂♂ | flesh-col. ♀♀ | flesh-col. ♂♂ | % black | % mosaic | % red | % flesh-col. | total ♀♀ | total ♂♂ |
|---|---|---|---|---|---|---|---|---|---|---|---|---|---|---|
| 116 B mos. F4* | | | 41 | 29 | 22 | 16 | | | | 64.8 | 35.2 | | 63 | 45 |
| F5* | | | 39 | 22 | 15 | 9 | | | | 71.7 | 28.3 | | 54 | 31 |
| F6* | 1 | 3 | 14 | 11 | 19 | 22 | | | 5.70 | 35.7 | 58.6 | | 34 | 36 |
| F7* | | | 17 | 14 | 1 | 4 | | | | 86.05 | 13.95 | | 18 | 18 |
| F8* | | | 25 | 26 | 18 | 20 | | | | 57.4 | 42.6 | | 43 | 46 |
| F9* | | | 15 | 17 | 24 | 32 | | | | 36.4 | 63.6 | | 39 | 49 |
| F10* | 4 | 5 | 22 | 20 | 5 | 5 | | | 14.78 | 68.8 | 16.4 | | 31 | 30 |
| F12* | | | 15 | 17 | 30 | 28 | 3 | | | 34.4 | 62.4 | 3.2 | 45 | 48 |
| F13 | | 1 | 26 | 18 | 29 | 15 | | | 1.2 | 49.4 | 49.4 | | 55 | 34 |
| 417 F1 | | | 33 | 38 | 21 | 24 | | | | 61.2 | 38.8 | | 54 | 62 |
| 424 AF1 | | 1 | 4 | 4 | 4 | 3 | 9 | | 4.0 | 32.0 | 28.0 | 36.0 | 8 | 17 |
| BF1 | | 1 | 34 | 37 | 11 | 14 | | | 1.0 | 73.2 | 25.8 | | 45 | 52 |
| CF1 | | | 12 | 9 | 10 | 23 | | | | 38.9 | 61.1 | | 22 | 32 |
| DF1 | 1 | | 39 | 35 | 16 | 21 | | | 0.9 | 66.1 | 33.0 | | 56 | 56 |
| GF1 | | | 2 | 1 | 3 | 3 | 5 | | | 21.4 | 42.8 | 35.8 | 5 | 9 |
| 417 AF2 | | | 13 | 6 | 5 | 14 | | | | 50.0 | 50.0 | | 18 | 20 |
| BF2 | | | 38 | 25 | 16 | 22 | | | | 62.4 | 37.6 | | 54 | 47 |
| GF2 | | | 87 | 85 | 106 | 110 | | | | 44.3 | 55.7 | | 193 | 195 |
| 424 AF2 | | | 24 | 31 | 14 | 20 | | | | 61.75 | 38.25 | | 38 | 51 |
| BF2 | | | 13 | 9 | 1 | 2 | | | | 88.0 | 12.0 | | 14 | 11 |
| CF2 | | | 22 | 34 | 34 | 25 | | | | 48.7 | 51.3 | | 56 | 59 |
| DF2 | | | 14 | 3 | 10 | 3 | 15 | | | 37.8 | 28.9 | 33.3 | 24 | 21 |
| EF2 | | | 18 | 24 | 12 | 18 | | | | 58.3 | 41.7 | | 30 | 42 |
| GF2 | | | 4 | | | 1 | | | | 80.0 | 20.0 | | 4 | 1 |
| HF2 | 4 | 2 | 42 | 39 | 20 | 25 | | | 4.5 | 61.4 | 34.1 | | 66 | 66 |
| V | | | 27 | 21 | 28 | 15 | 23 | | | 42.1 | 37.7 | 20.2 | 55 | 59 |
| total | 10 | 13 | 640 | 575 | 474 | 494 | 55 | | 1.0 | 53.7 | 42.8 | 2.5 | 1124 | 1137 |

\* Observations of ARENDSEN HEIN or FERWERDA.

Of 116 B mosaic the P-generation, the $F_1$, $F_2$ and $F_3$ have already been discussed above. They have not been included in table 5, because they were no crosses of mosaic-eyed individuals inter-se. From the $F_4$ up to and included the $F_{13}$ attempts have been made to produce a pure stock by continually propagating individuals with pronouncedly mosaic eyes, but without any result. In addition to mosaic-eyed individuals there always occurred red-eyed individuals, in a greatly variable number. Column 6 gives the percentage of black-eyed, mosaic-eyed, red-eyed and flesh-coloured individuals. From this it appears that the number of mosaic-eyed individuals may amount to upwards of 70% and may fall to about 30% of the total number of individuals in a generation. This points to the fact that in this case an ordinary Mendelian segregation is out of the question.

Of a total of 2261 individuals 53.7% was mosaic-eyed, 42.8% red-eyed, 1.0% black-eyed and 2.5% flesh-coloured. So there also arose a small number of black-eyed individuals. The occurrence of individuals with flesh-coloured eyes is probably a result of heterozygosis of some beetles. I shall revert to this later. It has not yet been possible to examine the black-eyed individuals for their genetic constitution, because they were nearly always deformed. Nevertheless a few black-eyed individuals have been crossed now; the result, however, is not yet known.

In addition to the selection of mosaic-eyed individuals, attempts have been made to apply selection in a different way, i.e. by inter-crossing individuals with eyes which were black only for a slight part and also individuals of which the eyes were predominantly black. The surface of the black part of the eyes compared with the total surface of the eyes was estimated and indicated by a fraction. If the eye colour of both eyes together was predominantly black, e.g. 2/3, the eye was called strongly mosaic, if the colour was predominantly red, e.g. 4/5, faintly mosaic. The strongly mosaic-eyed individuals were bred together, likewise the faintly mosaic-eyed ones. The eye colour of the offspring was determined and the degree of the mosaic ones was estimated, as indicated above. After this three grades were fixed, viz. smaller than 1/3, that is that the two eyes together were black for less than 1/3; from 1/3–2/3, in which the black part of the total surface of the eyes was at least 1/3 and at most 2/3, and larger than 2/3, with a black part larger than 2/3 of the total eye-surface. Between the two eyes there was as a rule little difference. In the two subjoined tables the degree of the mosaic character of the parents has been given in column 2.

Table 6 gives the result of the breeding of faintly mosaic-eyed beetles, table 7 that of the progenies of strongly mosaic-eyed ones.

TABLE 6. OFFSPRING OF FAINTLY MOSAIC INDIVIDUALS

| number | grade of the parents ♀ | grade of the parents ♂ | red ♀♀ | red ♂♂ | mosaic up to-⅓ ♀♀ | mosaic up to-⅓ ♂♂ | mosaic ⅓-⅔ ♀♀ | mosaic ⅓-⅔ ♂♂ | mosaic ⅔-entirely ♀♀ | mosaic ⅔-entirely ♂♂ | total ♀♀ | total ♂♂ |
|---|---|---|---|---|---|---|---|---|---|---|---|---|
| 424 B | ¼ | ⅙ | 1 | 2 | 7 | 1 | 4 | 3 | 2 | 5 | 14 | 11 |
| C | ¼ | ⅕ | 34 | 25 | 16 | 25 | 6 | 9 | | | 56 | 59 |
| E | ¼ | ⅙ | 12 | 18 | 5 | 10 | 8 | 7 | 5 | 7 | 30 | 42 |
| total | | | 47 | 45 | 28 | 36 | 18 | 19 | 7 | 12 | 100 | 112 |

TABLE 7. OFFSPRING OF STRONGLY MOSAIC INDIVIDUALS

| number | grade of the parents ♀ | grade of the parents ♂ | red ♀♀ | red ♂♂ | mosaic up to ⅓ ♀♀ | mosaic up to ⅓ ♂♂ | mosaic ⅓-⅔ ♀♀ | mosaic ⅓-⅔ ♂♂ | mosaic ⅔-entirely ♀♀ | mosaic ⅔-entirely ♂♂ | black ♀♀ | black ♂♂ | flesh-col. ♀♀ | flesh-col. ♂♂ | total ♀♀ | total ♂♂ |
|---|---|---|---|---|---|---|---|---|---|---|---|---|---|---|---|---|
| 424 A | ⅔ | ⅔ | 14 | 20 | | 6 | 9 | 15 | 15 | 10 | | | | | 38 | 51 |
| D | ⅔ | ⅔ | 10 | 3 | | | 7 | 1 | 7 | 2 | | | | 15 | 24 | 21 |
| G | ½ | ⅔ | | 1 | 1 | | 2 | | 1 | | | | | | 4 | 1 |
| H | ⅘ | ⅘ | 20 | 25 | 3 | 7 | 13 | 25 | 26 | 7 | 4 | 2 | | | 66 | 66 |
| total | | | 44 | 49 | 4 | 13 | 31 | 41 | 49 | 19 | 4 | 2 | | 15 | 132 | 139 |

Of table 6 all 3 were pair-matings; of table 7 424 A, D and G were pair-matings whereas 424 H was a cross between two ♀♀ and two ♂♂.

The totals of the tables 6 and 7 have been given in table 8, expressed in per cents.

TABLE 8. TOTALS OF TABLES 6 AND 7 IN PERCENTAGES

| | red | mosaic up to ⅓ | mosaic ⅓-⅔ | mosaic ⅔-entirely | black | flesh-col. |
|---|---|---|---|---|---|---|
| total of table 6 | 43.39 | 30.19 | 17.45 | 8.96 | | |
| total of table 7 | 34.32 | 6.27 | 26.57 | 25.09 | 2.21 | 5.53 |
| $m_{diff.}$ | 4.45 | 3.48 | 3.74 | 3.28 | | |
| $D/m_{diff.}$ | 2.04 | 6.87 | 2.44 | 4.91 | | |

From this table it may be concluded that there is no significant difference between the numbers of red-eyed individuals in table 6 and table 7, that there is likewise no significant difference between the numbers of beetles, of which the eyes were black for 1/3–2/3. Moreover in table 6 more beetles occur belonging to the group up to 1/3, on the other hand in table 7 more that belong to the group 2/3 to entirely. In addition black-eyed individuals only arose in the cross of strongly mosaic-eyed beetles. The selection, therefore, has indeed resulted in a shifting. In the following generation the selection has been continued.

The results of cross-breeding faintly and strongly mosaic-eyed beetles are not yet known.

We have been confronted by a remarkable fact in our attempts to obtain a melanistic stock with mosaic-eyes. We started with crossing mosaic-eyed dark brown beetles with red-eyed heterozygous melanistic ones. Homozygous melanistic beetles were not at our disposal at the moment. The result has been given in table 9.

TABLE 9. OFFSPRING OF MOSAIC-EYED, DARK BROWN × RED-EYED, HETEROZYGOUS MELANISTIC

|  | heterozygous melanistic | | dark brown | |
| --- | --- | --- | --- | --- |
|  | mosaic | red | mosaic | red |
| ♀♀ | 12 | 10 | 20 | 15 |
| ♂♂ | 16 | 10 | 10 | 13 |
| total | 28 | 20 | 30 | 28 |

The ratio melanistic : dark brown is 48 : 58, the ratio mosaic : red just the reverse, viz. 58 : 48.

The heterozygous melanistic mosaic-eyed individuals were bred together. Therefore homozygous melanistic mosaic-eyed beetles might theoretically also be expected in the progeny. The result, however, was different, like demonstrated by table 10.

TABLE 10. OFFSPRING OF MOSAIC-EYED, HETEROZYGOUS MELANISTIC BEETLES

| | homozygous melanistic | | heterozygous melanistic | | dark brown | |
|---|---|---|---|---|---|---|
| | mosaic | red | mosaic | red | mosaic | red |
| ♀♀ | | 12 | 23 | 21 | 14 | 9 |
| ♂♂ | | 6 | 25 | 22 | 11 | 5 |
| total | | 18 | 48 | 43 | 25 | 14 |

From this it appears that not a single homozygous melanistic mosaic-eyed individual has arisen in a total of 148 individuals. The number of homozygous melanistic individuals found, viz. 18, deviates greatly from the number of 37, which might theoretically be expected on a total number of 148 individuals. The deviation is significant: m being 5.27, $D/m = 3.60$. The only conclusion which may be drawn from this is, that the genotype homozygous melanistic with mosaic eyes is not viable, which conclusion is corroborated by the observations in the following generation, which has likewise been bred from heterozygous melanistic mosaic-eyed beetles, table 11.

TABLE 11. OFFSPRING OF MOSAIC-EYED, HETEROZYGOUS MELANISTIC BEETLES

| | homozygous melanistic | | heterozygous melanistic | | dark brown | |
|---|---|---|---|---|---|---|
| | mosaic | red | mosaic | red | mosaic | red |
| ♀♀ | | 15 | 37 | 21 | 26 | 18 |
| ♂♂ | 1 | 10 | 24 | 27 | 19 | 5 |
| total | 1 | 25 | 61 | 48 | 45 | 23 |

The only melanistic mosaic-eyed individual that arose was greatly deformed and consequently unsuitable for propagating purposes. For the rest this generation showed the same as the previous, as here too the ratio melanistic : heterozygous melanistic deviated. In this case it is 26 : 109. The deviation from the theoretically expected number of homozygous melanistic individuals, viz. 50.75 is significant: m = 6.17, D/m = 4.01. Indeed only half the number have arisen. Just now it is not yet possible for me to account for this. To be sure in this and in the previous generation an excess of ♀♀ occurred, the two generations together consisted of 196 ♀♀ and 155 ♂♂, but this does not account for the non-occurrence of the homozygous melanistic mosaic-eyed individuals, for of these both the ♀♀ and the ♂♂ failed to appear. This therefore is a secondary phenomenon, which if maintained in the following generations needs a special explanation.

As appears from the above red-eyed individuals were always found in the progeny of mosaic-eyed beetles. These red-eyed individuals were bred together and gave the following result, table 12.

TABLE 12. OFFSPRING OF RED-EYED BEETLES

| number | mosaic | | red | | total | |
|---|---|---|---|---|---|---|
| | ♀♀ | ♂♂ | ♀♀ | ♂♂ | ♀♀ | ♂♂ |
| 116 B red F12 | 2 | | 71 | 64 | 73 | 64 |
| F13 | 1 | 1 | 37 | 43 | 38 | 44 |
| 116 CF1 | | 1 | 29 | 37 | 29 | 38 |
| total | 3 | 2 | 137 | 144 | 140 | 146 |

In addition a progeny was bred from the red-eyed individuals arisen from crosses between mosaic-eyed and red-eyed beetles. The following result was obtained:

TABLE 13. OFFSPRING OF RED-EYED PROGENY FROM THE CROSS
MOSAIC ∾ RED

| number | mosaic | | red | | total | |
|---|---|---|---|---|---|---|
| | ♀♀ | ♂♂ | ♀♀ | ♂♂ | ♀♀ | ♂♂ |
| 352 F2 | | | 43 | 56 | 43 | 56 |
| 369 BF2 | 5 | 2 | 54 | 54 | 59 | 56 |
| 352 F3 | | 1 | 33 | 28 | 33 | 29 |
| 354 F2 | | | 61 | 56 | 61 | 56 |
| 359 CF2 | | | 7 | 3 | 7 | 3 |
| 367 BF2 | 8 | 7 | 53 | 44 | 61 | 51 |
| 420 AF2 | 3 | 1 | 7 | 3 | 10 | 4 |
| 358 BF3 | 1 | | 46 | 36 | 47 | 36 |
| total | 17 | 11 | 304 | 280 | 321 | 291 |

Finally a number of crosses have been made between purely red-eyed individuals and red-eyed $F_1$ individuals, arisen from crosses between mosaic-eyed ♀♀ and ♂♂ of a different eye colour (black, red or yellow). In these crosses therefore either only the females (359 D) or only the males (419 C, 419 D and 421 J) were descended from mosaic-eyed individuals. The result has been comprised in the table 14.

TABLE 14. RED (MOSAIC × RED) ∽ RED

| number | red | | black | | total | |
|---|---|---|---|---|---|---|
| | ♀♀ | ♂♂ | ♀♀ | ♂♂ | ♀♀ | ♂♂ |
| 359 D | 18 | 10 | | | 18 | 10 |
| 419 C | 81 | 75 | 3 | 2 | 84 | 77 |
| 419 D | 44 | 48 | 2 | | 46 | 48 |
| 421 J | 37 | 45 | | | 37 | 45 |
| total | 180 | 178 | 5 | 2 | 185 | 180 |

From a comparison of table 5 and table 12 it appears that from mosaic-eyed descendants from mosaic-eyed parents a great many more mosaic-eyed individuals arise than occur in the progeny of their red-eyed brothers and sisters inter-se.

TABLE 15. COMPARISON OF TOTALS OF TABLES 5 AND 12
IN PERCENTAGES

| | black | mosaic | red | flesh-col. |
|---|---|---|---|---|
| Table 5 | 1.0 | 53.7 | 42.8 | 2.5 |
| Table 12 | | 1.75 | 98.25 | |

The crosses comprised in table 13 can best be compared with those between mosaic-eyed individuals, which are themselves descendants from mosaic-eyed individuals crossed with red-eyed ones. These latter have been comprised in table 16; table 17 demonstrates the difference.

TABLE 16. MOSAIC × MOSAIC (FROM CROSSES BETWEEN MOSAIC ∽ RED)

| number | black | | mosaic | | red | |
|---|---|---|---|---|---|---|
| | ♀♀ | ♂♂ | ♀♀ | ♂♂ | ♀♀ | ♂♂ |
| 406 H | 4 | | 19 | 14 | 19 | 12 |
| 358 C | | | 9 | 15 | 4 | 10 |
| 367 A | | | 18 | 14 | 34 | 38 |
| 369 A | | | 11 | 27 | 30 | 40 |
| 376 A | | | 25 | 24 | 25 | 22 |
| 420 B | | | 26 | 10 | 26 | 19 |
| total | 4 | | 108 | 104 | 138 | 141 |

TABLE 17. COMPARISON OF TOTALS OF TABLES 13 AND 16 IN PERCENTAGES

| number | black | | mosaic | | red | |
|---|---|---|---|---|---|---|
| | ♀♀ | ♂♂ | ♀♀ | ♂♂ | ♀♀ | ♂♂ |
| Table 13 | | | 4.14 | | 95.86 | |
| Table 16 | 0.81 | | 42.83 | | 56.36 | |

From this it also appears that there is a great difference between the number of mosaic-eyed and red-eyed individuals arisen from crosses in which mosaic-eyed individuals are involved.

To be sure red-eyed descendants from mosaic-eyed individuals also yield a number of mosaic-eyed offspring, but this number is much smaller than in the cross of mosaic-eyed individuals inter-se. As opposed to this FERWERDA stated (p. 100) ,,Dabei zeigte sich mir, dass unter der Nachkommenschaft der rotäugigen Käfer auch wieder geflecktäugige Individuen vorkamen, ihre Anzahl war nicht geringer als bei der Nachkommenschaft der gefleckten Käfer." Probably he refers here to a cross made by him (116 B red $F_{10}$) of similar red-eyed individuals which yielded 22 ♀♀ and 20 ♂♂ mosaic-eyed and 5 ♀♀ + 5 ♂♂ red-eyed. ARENDSEN HEIN, however, had made the same cross before, from which arose: 1 mosaic-eyed female and 23 ♀♀ + 17 ♂♂ red-eyed (116 B red $F_9$), which result is in adrfect correspondence with my results.

On the whole it has therefore been ascertained that some, may be all red-eyed descendants from mosaic-eyed individuals are capable of producing again mosaic-eyed progeny, be it to a much smaller extent than the mosaic-eyed ones themselves.

### § 4. *Explanatory hypotheses*

The character mosaic eye impresses us strongly as that of an ever-sporting variety, or as it is often called nowadays an unstable character. Among other things it shows correspondence with the cases mentioned by DEMEREC (16, 1926; 17, 18, 1927) in *Drosophila simulans*, viz. reddish alpha, miniature alpha and magenta alpha.

It may be imagined that the mosaic-eyed individuals, just like the red-eyed ones have the genetic formula ffGGHH. The difference is then based upon the fact that the gene f in the mosaic-eyed individuals has repeatedly mutated to F, which does not take place in the red-eyed individuals. The black spots in the eyes arose, when such cells form part of the eye. In this case it must of course be assumed that each cell is self-differentiating, so that an eye-cell, in which the f has not mutated, will be red, the one in which mutation did take place black. When mutation does not occur in all cells mosaic-eyed individuals may arise. How the ratio between black and red will be, entirely depends on the number of mutations and the moment at which they occur.

About the question when these mutations should occur the following may be said. In the sexual cells, in the zygote and in the first cleavages no mutations occur or only very few. If this were the case, a great many more black-eyed individuals would arise than have occurred hitherto, assuming that the mutation f $\rightarrow$ F is not reversible. As in the development of the animals the eyes are late in differentiating, the greater part of the mutations will also take place late.

With the aid of this hypothesis all results hitherto discussed can be explained.

Besides the possibility that the occurrence of mosaic eyes is controlled by a mutable gene, there is an other possibility which likewise deserves attention. This refers to the origin of the first mosaic-eyed individuals, viz. in the $F_2$ of a cross between a red-eyed female and a black-eyed male. This cross yielded a black-eyed $F_1$. We can imagine that during the development of one of these $F_1$ individuals a translocation has occurred, owing to which a part of the chromosome containing the gene F was attached with this gene to the chromosome containing the gene f. This translocated part must have been present in some cells of the body of the individual in question, among others also in part of the sexual organs, with the result that this individual yielded the following gametes : F and f; in addition, however, also f with F attached, in future represented by $f_F$ and also o. Their ratio will depend on the part of the sexual organs in which the translocation occurred. Assuming that all these gametes were viable, the following possibilities of combination arose:

| gametes | F | f | $f_F$ | o |
|---------|-----|-----|--------|------|
| F | FF | Ff | $Ff_F$ | Fo |
| f | Ff | ff | $f_Ff$ | fo |

The FF, Ff, $Ff_F$ and Fo individuals will have been black-eyed, the ff and fo individuals red-eyed and as to the $f_Ff$ individuals we must assume that they have been the 7 mosaic-eyed individuals from the $F_2$.

It is known that in *Drosophila* translocated parts of chromosomes can be eliminated during the cell-divisions (PATTERSON, 61, 1932). If we assume that in the $f_Ff$ individuals this has also happened in a number of cell-divisions during the development, ff cells will have arisen in those individuals in addition to $f_Ff$ cells. Owing to the fact that the eyes contain the two kinds of cells they will have become partially black $f_Ff$, partially red ff.

Where the translocated F does not disappear, it must be assumed that the translocated part of chromosome has also divided because otherwise only eyes could arise which were black for at most one half.

From these $f_Ff$ individuals from the $F_2$ a great number of mosaic-eyed individuals has originated, both $f_Ff$ and $f_Ff_F$ individuals, from which again and again, possibly even in the reduction-divisions before the origin of the gametes, a number of translocated chromosome pieces have disappeared. Even because sooner or later in the development of the individuals an F may disappear there is little chance that black-eyed individuals will arise.

With this second hypothesis all results hitherto discussed can also be accounted for.

An objection to this hypothesis is, that it is hard to imagine that selection as described above, will have any result. Yet we should first await whether this selection is actually maintained in the following generations. If this is the case, this hypothesis can be abandoned as incorrect. In the first place, however, it is necessary to institute a cytological examination. Some attempts have been made in that direction, but up to now no good preparations have been obtained.

The two hypotheses mentioned are exclusively based on the results of cross-breeding mosaic-eyed individuals. Neither, however, can be correct, if not all results of cross-breeding can be explained by them. That is why a number of crosses have been made between mosaic-eyed individuals and individuals of a different eye colour. They will be discussed in succession, viz. mosaic $\infty$ black, mosaic $\infty$ red, mosaic $\infty$ yellow and mosaic $\infty$ flesh-coloured.

## § 5. *Crosses*

*a.* Mosaic $\infty$ black. In the above-mentioned hypotheses it has been assumed that the mosaic eye can only arise when the genetic formula of the individual in question is ffGGHH, i.e. the formula of the red-eyed form. From this it follows that the mosaic eye is a modified red. On the whole therefore mosaic will have to behave as if it were red, with the exception of the fact that a number of the ffGGHH individuals arisen in the crosses, will be mosaic-eyed.

In the results of the crosses of mosaic-eyed individuals and beetles of a different variety of eye colour, the red- and mosaic-eyed individuals will therefore have to be added up in their proportion to the other eye colour types that have arisen, as a result of the circumstance that the difference between red and mosaic is not a difference in gene, in which Mendelian segregations are to be expected.

$a_1$. $\female$ mosaic $\times$ $\male$ black. Quite according to expectations the $F_1$ consisted exclusively of black-eyed individuals, viz. 18 $\female\female$ and 18 $\male\male$. Mosaic is therefore recessive to black, which had already been discovered by FERWERDA. The black-eyed beetles from the $F_1$ bred together yielded the following offspring, table 18.

TABLE 18. $F_2$ OF MOSAIC × BLACK

| number | black | | mosaic | | red | | total | |
|---|---|---|---|---|---|---|---|---|
| | ♀♀ | ♂♂ | ♀♀ | ♂♂ | ♀♀ | ♂♂ | ♀♀ | ♂♂ |
| 418A | 37 | 36 | 4 | 1 | 8 | 2 | 49 | 39 |

|  | | | | |
|---|---|---|---|---|
| | 73 | | 15 | |
| theor. 3 : 1 | 66 | | 22 | |

$$m = 4.06 \quad D/m = 1.72.$$

The values found therefore do not depart considerably from the expected 3 : 1 ratio.

Black-eyed ♀♀ from the $F_1$ were back-crossed with mosaic-eyed ♂♂ with the following result:

TABLE 19. BLACK (MOSAIC × BLACK) × MOSAIC

| number | black | | mosaic | | red | | total | |
|---|---|---|---|---|---|---|---|---|
| | ♀♀ | ♂♂ | ♀♀ | ♂♂ | ♀♀ | ♂♂ | ♀♀ | ♂♂ |
| 418 C | 51 | 44 | 22 | 17 | 40 | 25 | 113 | 86 |

|  | | | | |
|---|---|---|---|---|
| | 95 | | 104 | |
| theor. 1 : 1 | 99.5 | | 99.5 | |

$$m = 7.10 \quad D/m = 0.634$$

Theoretically the reciprocal cross of black-eyed ♂♂ from the $F_1$ with mosaic-eyed females should likewise yield a 1 : 1 ratio. The values found correspond with this, table 20.

TABLE 20. MOSAIC × BLACK (MOSAIC × BLACK)

| number | black | | mosaic | | red | | total | |
|---|---|---|---|---|---|---|---|---|
| | ♀♀ | ♂♂ | ♀♀ | ♂♂ | ♀♀ | ♂♂ | ♀♀ | ♂♂ |
| 418 B | 22 | 26 | 13 | 15 | 9 | 16 | 44 | 57 |

|  | | | | |
|---|---|---|---|---|
| | 48 | | 53 | |
| theor. 1 : 1 | 50.5 | | 50.5 | |

$$m = 5.02 \quad D/m = 0.50$$

The cross between ♀ mosaic and ♂ black-eyed has once more been made unfortunately with heterozygous material, with the result that segregation already took place in the $F_1$, table 21.

TABLE 21. $F_1$ OF MOSAIC × BLACK

| number | black | | mosaic | | red | | yellow | | total | |
|---|---|---|---|---|---|---|---|---|---|---|
| | ♀♀ | ♂♂ | ♀♀ | ♂♂ | ♀♀ | ♂♂ | ♀♀ | ♂♂ | ♀♀ | ♂♂ |
| 419 | 9 | 6 | 3 | | 3 | 3 | | 8 | 15 | 17 |

Probably the formula of the ♀ has been ffGGHh, that of the male FfGGHo. From such a cross we may theoretically expect a ratio of 3 black : 3 (mosaic + red) : 2 yellow, viz. 12 : 12 : 8, m = 2.73, D/m = 1.10.

Matings of the black-eyed individuals from the $F_1$ together yielded a very complicated $F_2$, consisting of:

TABLE 22. $F_2$ OF MOSAIC × BLACK

| number | black | | mosaic | | red | | yellow | | flesh-col. | | total | |
|---|---|---|---|---|---|---|---|---|---|---|---|---|
| | ♀♀ | ♂♂ | ♀♀ | ♂♂ | ♀♀ | ♂♂ | ♀♀ | ♂♂ | ♀♀ | ♂♂ | ♀♀ | ♂♂ |
| 419 A | 34 | 16 | 6 | 2 | 18 | 8 | 21 | | 11 | | 58 | 58 |
| 419 B | 25 | 9 | 6 | 8 | 13 | 7 | 17 | | 10 | | 44 | 51 |
| total | 59 | 25 | 12 | 10 | 31 | 15 | 38 | | 21 | | 102 | 109 |

Probably an error has crept in here in consequence of there having been red-eyed individuals among the black-eyed ones, which, because they were already in an advanced stage of darkening, might easily be taken for black-eyed ones. The result in 419 B, which was a pair-mating, may be accounted for by assuming that one individual was practically red-eyed. The ratio to be expected is then 3 black : 3 (mosaic + red) : 2 yellow, or in a total of 95 individuals 35.625 : 35.625 : 23.75. The ratio found is 34 : 34 : 27; m is respectively 4.718 and 4.22, D/m = 0.34 and 0.89. 419 A, which was not a pair-mating, cannot but yield a very intricate ratio, which cannot be simply represented by a formula.

Red-eyed $F_1$ ♂♂ back-crossed with strange red-eyed ♀♀ of pure red-eyed stock yielded the following offspring.

TABLE 23. RED × RED (MOSAIC × BLACK)

| number | black | | red | | total | |
|---|---|---|---|---|---|---|
| | ♀♀ | ♂♂ | ♀♀ | ♂♂ | ♀♀ | ♂♂ |
| 419 C | 3 | 2 | 81 | 75 | 84 | 77 |
| 419 D | 2 | | 44 | 48 | 46 | 48 |
| total | 5 | 2 | 125 | 123 | 130 | 125 |

It is difficult to give a reasonable explanation of the origin of black-eyed individuals in these crosses. Mosaic-eyed ones might have arisen in such a number. They may have to be regarded as impurities. The origin of red-eyed beetles is quite according to expectation.

Red-eyed $F_1$ individuals bred together yielded a peculiar result:

TABLE 24. $F_2$ OF MOSAIC × BLACK

| | red | | flesh-coloured | | total | |
|---|---|---|---|---|---|---|
| | ♀♀ | ♂♂ | ♀♀ | ♂♂ | ♀♀ | ♂♂ |
| 419 E | 22 | 11 | | 11 | 22 | 22 |

This is exactly a ratio of 3 red : 1 flesh-coloured. The data on the eye colour furnished by FERWERDA are not sufficient to account for the fact that only flesh-coloured males have arisen. I shall revert to this later.

$a_2$. ♀ black × ♂ mosaic. The $F_1$ was equal to that of the reciprocal cross, viz. entirely black-eyed and consisted of 18 ♀♀ + 15 ♂♂.

The black-eyed $F_1$ individuals bred together yielded the following result:

TABLE 25. $F_2$ OF BLACK $\times$ MOSAIC

| number | black ♀♀ | black ♂♂ | mosaic ♀♀ | mosaic ♂♂ | red ♀♀ | red ♂♂ | total ♀♀ | total ♂♂ |
|---|---|---|---|---|---|---|---|---|
| 358 A | 127 | 101 | 20 | 10 | 28 | 22 | 175 | 133 |

|  | 228 | 80 |
|---|---|---|
| theor. 3 : 1 | 231 | 77 |

$$m = 7.6 \qquad D/m = 0.39$$

The deviations from the theoretically expected 3 : 1 ratio therefore are well within the limits of error.

A second cross between black-eyed ♀♀ and mosaic-eyed ♂♂ yielded:

TABLE 26. $F_1$ OF BLACK $\times$ MOSAIC

| number | black ♀♀ | black ♂♂ | mosaic ♀♀ | mosaic ♂♂ | red ♀♀ | red ♂♂ | total ♀♀ | total ♂♂ |
|---|---|---|---|---|---|---|---|---|
| 406 | 34 | 34 | 1 | 2 | 7 | 6 | 42 | 42 |

Part of the black-eyed ♀♀ were evidently heterozygous Ff, hence the occurrence of mosaic- and red-eyed individuals in the $F_1$.

The $F_2$ resulting from the breeding together of black-eyed $F_1$ individuals has been comprised in table 27.

TABLE 27. $F_2$ OF BLACK $\times$ MOSAIC

| number | black ♀♀ | black ♂♂ | mosaic ♀♀ | mosaic ♂♂ | red ♀♀ | red ♂♂ | yellow ♀♀ | yellow ♂♂ | total ♀♀ | total ♂♂ |
|---|---|---|---|---|---|---|---|---|---|---|
| 406 A | 38 | 50 |  | 2 | 8 | 7 | 6 | 2 | 52 | 61 |

I cannot account for this. I had expected a ratio of 3 black : 1 (mosaic + red); it is not clear to me how a number of yellow-eyed individuals could also arise.

The cross red × red from the $F_1$ gave a normal result of 34 ♀♀ + 30 ♂♂ all of them red-eyed.

Red-eyed $F_1$ ♀♀ × black-eyed $F_1$ ♂♂ should theoretically yield a ratio of 1 black : 1 (mosaic? + red). The result found is: 21 black : 11 red. The deviation is therefore rather large, but it may, in my opinion, be largely put down to the small figures.

Mosaic-eyed $F_1$ individuals, bred together yielded, as might be expected, a large percentage of mosaic-eyed offspring, table 28.

TABLE 28. $F_2$ OF BLACK × MOSAIC

| number | black | | mosaic | | red | | total | |
|--------|------|------|------|------|------|------|------|------|
| | ♀♀ | ♂♂ | ♀♀ | ♂♂ | ♀♀ | ♂♂ | ♀♀ | ♂♂ |
| 406 H | 4 | | 19 | 14 | 19 | 12 | 42 | 26 |

Finally both black-eyed ♂♂ and ♀♀ from the $F_1$ were back-crossed with mosaic-eyed beetles. In both crosses a ratio of 1 black : 1 (mosaic + red) could be expected. The result of the two crosses has been given in table 29 in the above-mentioned order.

TABLE 29. BACK-CROSSES OF BLACK $F_1$ WITH MOSAIC

| number | black | | mosaic | | red | | total | |
|--------|------|------|------|------|------|------|------|------|
| | ♀♀ | ♂♂ | ♀♀ | ♂♂ | ♀♀ | ♂♂ | ♀♀ | ♂♂ |
| 406 l | 9 | 9 | 4 | 5 | 4 | 3 | 17 | 17 |
| 406 K | 26 | 31 | 14 | 9 | 21 | 17 | 61 | 57 |
| total | 35 | 40 | 18 | 14 | 25 | 20 | 78 | 74 |
| | | 75 | | 77 | | | | |
| theor. 1 : 1 | | 76 | | 76 | | | | |

The result of the two is well up to expectations.

The general impression of this series of crosses therefore is favourable as regards the hypotheses assumed.

*b.* Mosaic ∽ red. In all of the following results of cross-breeding the theoretical expectation will likewise be based upon the two hypotheses discussed in § 4.

$b_1$. ♀ Mosaic × ♂ red. The $F_1$ of these crosses was heterogeneous.

TABLE 30. $F_1$ OF MOSAIC × RED

| number | black ♀♀ | black ♂♂ | mosaic ♀♀ | mosaic ♂♂ | red ♀♀ | red ♂♂ | flesh-col. ♀♀ | flesh-col. ♂♂ | total ♀♀ | total ♂♂ |
|---|---|---|---|---|---|---|---|---|---|---|
| 354 | | | | | 2 | 4 | | | 2 | 4 |
| 359 | | | 7 | 9 | 11 | 12 | | | 18 | 21 |
| 367 | 2 | | 14 | 14 | 15 | 29 | | | 31 | 43 |
| 376 | | | 13 | 13 | 11 | 16 | | | 24 | 29 |
| 420 | | | 13 | 6 | 9 | 12 | | 16 | 22 | 34 |
| total | 2 | | 47 | 42 | 48 | 73 | | 16 | 97 | 131 |

It is a striking fact that in these $F_1$ generations proportionately more mosaic-eyed individuals have originated than in the reciprocal ones. I doubt, however, that this should be due to maternal protoplasm; for among other things tables 37 and 38 prove nothing of the kind. It seems more probable to me that it should be due to differences in the starting material.

For the $F_2$ a number of mosaic-eyed beetles from the $F_1$ was intercrossed with the following result, table 31.

TABLE 31. $F_2$ OF MOSAIC × RED

| number | mosaic ♀♀ | mosaic ♂♂ | red ♀♀ | red ♂♂ | total ♀♀ | total ♂♂ |
|---|---|---|---|---|---|---|
| 367 A | 18 | 14 | 34 | 38 | 52 | 52 |
| 376 A | 25 | 24 | 25 | 22 | 50 | 46 |
| 420 B | 26 | 10 | 26 | 19 | 52 | 29 |
| total | 69 | 48 | 85 | 79 | 154 | 127 |

Expressed in percents this is 41.64% mosaic-eyed, 58.36% red-eyed. The percentage of mosaic-eyed individuals is therefore fairly high. For the rest the result corresponds very well with that of the cross between mosaic-eyed individuals discussed above. So I think I am permitted to refer to the explanation given there.

Breeding together red-eyed $F_1$ individuals has, as was expected, yielded a great excess of red-eyed progeny, table 32.

TABLE 32. $F_2$ OF MOSAIC × RED

| number | mosaic ♀♀ | mosaic ♂♂ | red ♀♀ | red ♂♂ | total ♀♀ | total ♂♂ |
|---|---|---|---|---|---|---|
| 354 | | | 61 | 56 | 61 | 56 |
| 359 C | | | 7 | 3 | 7 | 3 |
| 367 B | 8 | 7 | 53 | 44 | 61 | 51 |
| 420 A | 3 | 1 | 7 | 3 | 10 | 4 |
| total | 11 | 8 | 128 | 106 | 139 | 114 |

This is respectively 92.47% red-eyed and 7.51% mosaic-eyed.

The crosses between mosaic-eyed ♀♀ and red-eyed ♂♂ from the $F_1$ gave a result, mentioned in table 33.

TABLE 33. MOSAIC × RED (OUT OF THE $F_1$)

| number | mosaic ♀♀ | mosaic ♂♂ | red ♀♀ | red ♂♂ | flesh-col. ♀♀ | flesh-col. ♂♂ | total ♀♀ | total ♂♂ |
|---|---|---|---|---|---|---|---|---|
| 359 A | | | 8 | 19 | | | 8 | 19 |
| 420 D | 17 | 15 | 34 | 28 | 13 | 19 | 64 | 62 |
| total | 17 | 15 | 42 | 47 | 13 | 19 | 72 | 81 |

Of the individuals of the formula ffGGHH therefore 26.45% was mosaic-eyed and 73.55% red-eyed. Here we are struck by the fact that in 420 D only mosaic-eyed individuals arose and none in 359 A. Probably, however, this will be connected with the small number of individuals in the last-mentioned cross. For the rest the result cannot be called unsatisfactory.

The origin of individuals with flesh-coloured eyes must be accounted for by assuming that the mosaic- and red-eyed individuals from the $F_1$ had the formula ffGgHH, resulting in a ratio of 3 (mosaic + red) : 1 flesh-coloured. A ratio 94 : 32 was found.

Reciprocal crossing, three times repeated yielded the following result:

TABLE 34. RED × MOSAIC (OUT OF THE $F_1$)

| number | black | | mosaic | | red | | total | |
|---|---|---|---|---|---|---|---|---|
| | ♀♀ | ♂♂ | ♀♀ | ♂♂ | ♀♀ | ♂♂ | ♀♀ | ♂♂ |
| 359 B | | 1 | 3 | | 7 | 4 | 10 | 5 |
| 376 B | | | 6 | 13 | 13 | 12 | i9 | 25 |
| 420 C | 1 | | | | 48 | 28 | 49 | 28 |
| total | 1 | 1 | 9 | 13 | 68 | 44 | 78 | 58 |

The first two show a tendency to uniformity, 420 C, however, is quite different, because in a total of 76 ffGGHH individuals not a single mosaic-eyed one has arisen. I had expected that about 20% of mosaic-eyed individuals would arise.

Of less importance is finally the cross between a red-eyed $F_1$ ♀ and a strange red-eyed ♂. It yielded 18 ♀♀ + 10 ♂♂, all red-eyed, as was expected.

$b_2$. ♀ red × ♂ mosaic. Three crosses have been made between red-eyed females and mosaic-eyed males with the following result:

TABLE 35. $F_1$ OF RED × MOSAIC

| number | mosaic | | red | | total | |
|---|---|---|---|---|---|---|
| | ♀♀ | ♂♂ | ♀♀ | ♂♂ | ♀♀ | ♂♂ |
| 352 | | | 11 | 22 | 11 | 22 |
| 369 | 13 | 18 | 25 | 28 | 38 | 46 |
| 433 | 21 | 16 | 93 | 91 | 114 | 107 |
| total | 34 | 34 | 129 | 141 | 163 | 175 |

68 mosaic-eyed individuals arose on a total of 338, that is 20.12%, a little less therefore than half the number of mosaic-eyed individuals resulting from the breeding together of mosaic-eyed beetles. This agrees very well with what might be theoretically expected, because only half the number of beetles used were also mosaic-eyed.

The red-eyed individuals from the $F_1$, which were bred together, yielded entirely according to expectation an excess of red-eyed progeny, in addition to some mosaic-eyed ones, respectively 97 ♀♀ + 110 ♂♂ and 5 ♀♀ + 2 ♂♂ or 96,73% red-eyed and 3.27% mosaic-eyed individuals. This result is fairly equal to that mentioned above, table 32.

The mosaic-eyed beetles from the $F_1$ were likewise bred together. This produced an $F_2$, consisting of:

TABLE 36. MOSAIC × MOSAIC (OUT OF THE $F_1$)

| number | mosaic | | red | | total | |
|---|---|---|---|---|---|---|
| | ♀♀ | ♂♂ | ♀♀ | ♂♂ | ♀♀ | ♂♂ |
| 369 A | 11 | 27 | 30 | 40 | 41 | 67 |

This is respectively 35.19% mosaic-eyed and 64.81% red-eyed. It cannot be exactly stated how large the number of mosaic-eyed individuals is, that might be expected theoretically, because it is not known how often the gene f mutates to F eventually the translocated gene F disappears. Indeed it might be expected that this number would be larger than that in the $F_1$ and also that it would be lower than the number of mosaic-eyed individuals, resulting from breeding together mosaic-eyed individuals. The beetles used for this cross have had at most one chromosome in which the mutable gene occurred or to which the translocated gene F must have been attached. The chromosome coming from the father was normal. Of course this influences the number of mosaic-eyed individuals which may arise from such a cross, hence the expectation expressed above.

Besides the two crosses discussed, two more crosses were made between red- and mosaic-eyed beetles from the $F_1$ t.w. two reciprocal crosses. Red × mosaic gave as its result:

TABLE 37. RED × MOSAIC (OUT OF THE $F_1$)

| number | mosaic | | red | | total | |
|---|---|---|---|---|---|---|
| | ♀♀ | ♂♂ | ♀♀ | ♂♂ | ♀♀ | ♂♂ |
| 369 C | 11 | 16 | 28 | 31 | 39 | 47 |

This is 31.39% mosaic- and 68.61% red-eyed. The number of mosaic-eyed individuals is a little higher than I had expected, an excess of mosaic-eyed individuals there is in my opinion, however, not. The reciprocal cross yielded only a slight number of $F_2$ individuals. On a total of 43 individuals I found:

TABLE 38. MOSAIC × RED (OUT OF THE $F_2$)

| number | mosaic | | red | | total | |
|---|---|---|---|---|---|---|
| | ♀♀ | ♂♂ | ♀♀ | ♂♂ | ♀♀ | ♂♂ |
| 369 D | 6 | 1 | 16 | 20 | 22 | 21 |

This is respectively 16.28% and 83.72%.

The number of mosaic-eyed individuals therefore is in this cross much smaller than in the preceding one, and more according to expectation. I think, however, I should not attach too much value to this difference in view of the small numbers.

From the $F_2$ a number of red-eyed beetles were bred together. The individuals used for it were offspring of red-eyed beetles from the $F_1$. They yielded an $F_3$ consisting of 33 ♀♀ and 28 ♂♂, all red-eyed and 1 mosaic-eyed ♂. This result is in perfect correspondence with the result of the crosses given in table 32.

*c.* Mosaic ∽ yellow.

$c_1$. ♀ mosaic × ♂ yellow.

As might be expected the $F_1$ consisted of black-eyed individuals only, viz. 37 ♀♀ + 55 ♂♂. These were bred together and yielded:

TABLE 39. $F_2$ OF MOSAIC × YELLOW

| number | black | | red | | yellow | | total | |
|---|---|---|---|---|---|---|---|---|
| | ♀♀ | ♂♂ | ♀♀ | ♂♂ | ♀♀ | ♂♂ | ♀♀ | ♂♂ |
| 405 A | 55 | 38 | 22 | 10 | | 40 | 77 | 88 |
| | 93 | | 32 | | 40 | | | |
| theor. 9 : 3 : 4 | 92.81 | | 30.94 | | 41.25 | | | |
| m | 6.37 | | 5.01 | | 5.56 | | | |
| D/m | 0.03 | | 0.21 | | 0.22 | | | |

The ratio found is well in accordance with the ratio 9 : 3 : 4 to be expected.

One black-eyed ♀ from the $F_1$ was back-crossed with one yellow-eyed ♂. A 1 : 1 ratio of black and yellow was to be expected. But red-eyed individuals also originated (see table 40). The explanation will have to be looked for in the fact that the ♂ used was heterozygous Ff. The formula of this individual would then have been FfGGho which is quite possible, as both FFGGhh and ffGGhh are yellow-eyed. In this case the theoretically expected ratio would be : 3 black : 1 red : 4 yellow, which is well in accordance with the values found.

TABLE 40. BACK-CROSS OF BLACK $F_1$ × YELLOW

| number | black ♀♀ | black ♂♂ | red ♀♀ | red ♂♂ | yellow ♀♀ | yellow ♂♂ | total ♀♀ | total ♂♂ |
|---|---|---|---|---|---|---|---|---|
| 405 B | 8 | 11 | 4 | 2 | 15 | 12 | 27 | 25 |
| | 19 | | 6 | | 27 | | | |
| theor. 3 : 1 : 4 | 19.5 | | 6.5 | | 26 | | | |

It is a remarkable fact that in this and in the preceding cross among the ffGGHH individuals not a single mosaic-eyed individual was found.

Back-crossing black-eyed ♂♂ from the $F_1$ with yellow-eyed ♀♀ yielded the result mentioned in table 41.

TABLE 41. BACK-CROSS OF YELLOW × BLACK $F_1$

| number | black ♀♀ ♂♂ | mosaic ♀♀ ♂♂ | red ♀♀ ♂♂ | yellow ♀♀ ♂♂ | total ♀♀ ♂♂ |
|---|---|---|---|---|---|
| 405 D | 31 | 5 | 13 | 2    52 | 51    52 |
| | 31 | 18 | | 52 | |
| theor. 3 : 1 : 4 | 37.88 | 12.62 | | 50.5 | |
| | m = 4.86 | m = 3.32 | | m = 5.02 | |
| | D/m = 1.42 | D/m = 1.62 | | D/m = 0.3 | |

It is probable that also in this cross the females which originated from the same stock as the ♀ from the preceding cross, had the formula FfGGhh. The 3 : 1 : 4 ratio to be expected in this case is well in accordance with the values found.

Back-crosses have also been made with mosaic-eyed individuals. The cross black-eyed ♀♀ from the $F_1$ with mosaic-eyed ♂♂ yielded an $F_2$, of which it was expected that it would consist of black-eyed, (mosaic + red-eyed) and yellow-eyed individuals in the ratio 3 : 3 : 2.

TABLE 42. BACK-CROSSES OF BLACK $F_1$ × MOSAIC

| number | black ♀♀ | black ♂♂ | mosaic ♀♀ | mosaic ♂♂ | red ♀♀ | red ♂♂ | yellow ♀♀ ♂♂ | total ♀♀ | total ♂♂ |
|---|---|---|---|---|---|---|---|---|---|
| 405 H | 41 | 17 | 29 | 3 | 27 | 21 | 39 | 97 | 80 |
| 421 E | 94 | 46 | 34 | 11 | 47 | 37 | 84 | 175 | 178 |
| total | 135 | 63 | 63 | 14 | 74 | 58 | 123 | 272 | 258 |

|  | 198 | 209 | 123 |
|---|---|---|---|
| theor. 3 : 3 : 2 | 198.75 | 198.75 | 132.5 |
|  | m = 11.14 | m = 11.14 | m = 9.96 |
|  | D/m = 0.067 | D/m = 0.92 | D/m = 0.95 |

The correspondence is therefore quite satisfactory.

Black-eyed ♂♂ back-crossed with mosaic-eyed ♀♀ yielded an $F_2$, which was expected to consist of black-eyed and (mosaic-eyed + red-eyed) individuals in the ratio 1 : 1. The findings, however, did not quite answer to this, table 43.

TABLE 43. BACK-CROSSES of MOSAIC × BLACK $F_1$

| number | black ♀♀ | black ♂♂ | mosaic ♀♀ | mosaic ♂♂ | red ♀♀ | red ♂♂ | yellow ♀♀ ♂♂ | flesh-col. ♀♀ | flesh-col. ♂♂ | total ♀♀ | total ♂♂ |
|---|---|---|---|---|---|---|---|---|---|---|---|
| 405 E | 23 | 20 | 11 | 5 | 11 | 18 | 6 | 3 | 5 | 48 | 54 |
| 421 G | 34 | 31 | 10 | 9 | 18 | 25 | 1 | | | 62 | 66 |
| total | 57 | 51 | 21 | 14 | 29 | 43 | 7 | 3 | 5 | 110 | 120 |

|  | 108 | 107 |
|---|---|---|

I cannot account for the fact that in this cross in addition to yellow-eyed ♂♂ also flesh-coloured individuals arose. Presumably this is due to the intruding of strange individuals in 405 E.

Possibly the yellow-eyed ♂ in 421 G is a result of non-disjunction.

Hitherto I have discussed of 421 only those crosses which came up plainly to expectation. The more complicated results may follow now. The $F_1$ consisted of:

TABLE 44. $F_1$ OF MOSAIC × YELLOW

| number | black | | mosaic | | red | | total | |
|--------|-------|---|--------|---|-----|---|-------|---|
|        | ♀♀    | ♂♂ | ♀♀    | ♂♂ | ♀♀ | ♂♂ | ♀♀   | ♂♂ |
| 421    | 25    | 40 | 6     | 5  | 5  | 8  | 36   | 53 |

At any rate one of the parents has evidently been heterozygous in this case. The result may be accounted for by assuming that one of the ♂♂ used was heterozygous Ff, the other homozygous FF.

From the $F_1$ two crosses have been made between black-eyed individuals, table 45.

TABLE 45. $F_2$ OF MOSAIC × YELLOW

| number | black | | mosaic | | red | | yellow | | flesh-col. | | total | |
|--------|-------|----|--------|---|-----|----|--------|----|------------|---|-------|-----|
|        | ♀♀    | ♂♂ | ♀♀    | ♂♂ | ♀♀ | ♂♂ | ♀♀    | ♂♂ | ♀♀        | ♂♂ | ♀♀   | ♂♂ |
| 421 A  | 9     | 7  | 2      | 2 | 6  | 3  | 2.     | 8  |           | 3  | 19    | 23  |
| 421 D  | 98    | 40 | 10     | 3 | 26 | 12 | 2?     | 57 |           | 16 | 136   | 128 |
| total  | 107   | 47 | 12     | 5 | 32 | 15 | 4?     | 65 |           | 19 | 155   | 151 |

|         | 154      | 64      | 84      |
|---------|----------|---------|---------|
| theor.  |          |         |         |
| 9 : 3 : 4 | 169.875 | 56.625 | 75.5 |
| m =     | 8.62     | 6.78    | 7.52    |
| D/m =   | 1.84     | 1.09    | 1.13    |

This is much more intricate than the 9 : 3 : 4 ratio expected. It is, however, possible to introduce a simplification by adding the individuals recorded as flesh-coloured to the yellow-eyed ones. In my opinion there is a reason for doing so, though initially the individuals were recorded as flesh-coloured. As contrasted with the flesh-coloured eye the yellow eye grows red after some time. This process of reddening lasts shorter than 7 days in the incubator. During the observations individuals of which I was not sure whether they were yellow or flesh-coloured, were placed into the incubator and after at least 7 days they were examined again. Individuals of which the eyes were not yet red, were recorded as flesh-coloured. However at the moment I am writing this, there is another $F_2$ of the cross between mosaic-eyed and yellow-eyed individuals that has to be examined. It has now appeared to me that after 7 days some individuals still have entirely unchanged eyes, whereas after a stay of 18 days in the incubator those eyes have grown red. In the $F_2$ of such a cross individuals of the formulae FF(f)GGhh and ffGGhh will arise side by side in a 3 : 1 ratio. My supposition is, that it are the latter individuals which darken so slowly and this supposition is supported by the figures found in 421 A and 421 D, which give an almost exact 3 : 1 ratio between the individuals recorded as yellow-eyed and those recorded as flesh-coloured. The proof that the „flesh-coloured" individuals indeed are yellow-eyed ffGGho was furnished by crossing them with individuals of the pure flesh-coloured strain. Also the in table 5 as flesh-coloured recorded individuals very probably have been yellow-eyed. This also agrees very well with the ratio found in those crosses between mosaic + red on the one side and flesh-coloured on the other, table 46.

TABLE 46. OFFSPRING OF MOSAIC BEETLES

| number | black ♀♀ | black ♂♂ | mosaic ♀♀ | mosaic ♂♂ | red ♀♀ | red ♂♂ | flesh-col. ♀♀ | flesh-col. ♂♂ | total ♀♀ | total ♂♂ |
|---|---|---|---|---|---|---|---|---|---|---|
| 424 AF1 | | 1 | 4 | 4 | 4 | 3 | | 9 | 8 | 17 |
| 424 GF1 | | | 2 | 1 | 3 | 3 | | 5 | 5 | 9 |
| 424 DF2 | | | 14 | 3 | 10 | 3 | | 15 | 24 | 21 |
| V | | | 27 | 21 | 28 | 15 | | 23 | 55 | 59 |
| total | | 1 | 47 | 29 | 45 | 24 | | 52 | 92 | 106 |

146                52
theor. 3 : 1                148.5                49.5
m = 6.09        D/m = 0.41

I suppose that the individuals used in these crosses have had the formulae ffFFHH and ffGGHo, which gave rise to the said 3 : 1 ratio in the progeny. The yellow-eyed individuals can only have the formula ffGGhh and this accounts at the same time for the fact that only „flesh-coloured" males originated.

We now revert to 421 $AF_2$ and 421 $DF_2$. Assuming the above supposition to be correct, we get a ratio of 154 black : 64 (mosaic + red) : 84 yellow. Theoretically it should be 169.875 black : 56.625 (mosaic + red) : 75.5 yellow. m is respectively = 8.62, 6.78, 7.52. D/m = 1.84, 1.09 and 1.13. The deviations therefore lie within the limits of error.

As already discussed above a ratio 1 black : 1 yellow might be expected in the back-cross of black-eyed ♀♀ from the $F_1$ with yellow-eyed ♂♂. The result was an $F_2$ consisting of 55 ♀♀ + 57 ♂♂ black-eyed and 68 ♀♀ + 91 ♂♂ yellow-eyed. Though the expectation that only black- and yellow-eyed individuals would arise, has been justified, the ratio is about 1 : 1.5. I cannot account for the cause of this. At any rate it is better in the reciprocal cross which yielded 77 black-eyed ♀♀ and 74 yellow-eyed ♂♂.

$c_2$. ♀ yellow × ♂ mosaic. As might be expected the $F_1$ consisted of black-eyed ♀♀ and yellow-eyed ♂♂.

TABLE 47. $F_1$ OF YELLOW × MOSAIC

| number | black | | yellow | | total | |
|---|---|---|---|---|---|---|
| | ♀♀ | ♂♂ | ♀♀ | ♂♂ | ♀♀ | ♂♂ |
| 370 | 63 | | | 53 | 63 | 53 |
| 432 A | 93 | | | 94 | 93 | 94 |
| 432 B | 47 | 1 | | 54 | 47 | 55 |
| total | 203 | 1 | | 201 | 203 | 202 |

The black-eyed ♂ will probably have resulted from non-disjunction. Black-eyed ♀♀ from the $F_1$ were crossed with yellow-eyed ♂♂ likewise from the $F_1$ with the following result:

TABLE 48. BLACK $F_1$ × YELLOW $F_1$

| number | black ♀♀ | black ♂♂ | mosaic ♀♀ | mosaic ♂♂ | red ♀♀ | red ♂♂ | yellow ♀♀ | yellow ♂♂ | total ♀♀ | total ♂♂ |
|---|---|---|---|---|---|---|---|---|---|---|
| 370 A | 24 | 35 | 8 | 3 | 6 | 8 | 39 | 47 | 77 | 93 |
| 432 A | 65 | 58 | 4 | 8 | 13 | 16 | 96 | 83 | 178 | 165 |
| total | 89 | 93 | 12 | 11 | 19 | 24 | 135 | 130 | 255 | 258 |
| | 182 | | 66 | | | | 265 | | | |
| theor. 3 : 1 : 4 | 192.375 | | 64.125 | | | | 256.5 | | | |
| m | 10.97 | | 7.49 | | | | 11.32 | | | |
| D/m | 0.95 | | 0.25 | | | | 0.75 | | | |

The values found therefore show a satisfactory correspondence with the 3 : 1 : 4 ratio expected.

Black-eyed ♀♀ from the $F_1$ back-crossed with yellow-eyed ♂♂ should give an $F_2$ consisting of black-eyed and yellow-eyed individuals in the ratio 1 : 1. The results correspond with this, table 49.

TABLE 49. BACK-CROSS: BLACK $F_1$ × YELLOW

| number | black ♀♀ | black ♂♂ | yellow ♀♀ | yellow ♂♂ | total ♀♀ | total ♂♂ |
|---|---|---|---|---|---|---|
| 370 C | 27 | 36 | 33 | 25 | 60 | 61 |
| | 63 | | 58 | | | |
| theor. 1 : 1 | 60.5 | | 60.5 | | | |
| | m = 5.5. | | D/m = 0.45 | | | |

Yellow-eyed ♂♂ from the $F_1$, back-crossed with yellow-eyed ♀♀ yielded, as was expected exclusively yellow-eyed progeny, viz. 72 ♀♀ + 46 ♂♂. The great difference in number of ♀♀ and ♂♂ is remarkable in this case.

Back-crossing yellow-eyed ♂♂ from the $F_1$ with mosaic-eyed ♀♀ has proved a failure.

Finally some black-eyed ♀♀ from the $F_1$ have been back-crossed with mosaic-eyed ♂♂.

TABLE 50. BACK-CROSS: BLACK $F_1$ × MOSAIC

| number | black<br>♀♀   ♂♂ | | mosaic<br>♀♀   ♂♂ | | red<br>♀♀   ♂♂ | | yellow<br>♀♀   ♂♂ | | total<br>♀♀   ♂♂ | |
|---|---|---|---|---|---|---|---|---|---|---|
| 370 E | 16 | 7 | 3 | 5 | 4 | 8 | 7 | 12 | 30 | 32 |
| | 23 | | 20 | | | | 12 | | | |
| theor. 3 : 3 : 2 | 20.625 | | 20.625 | | | | 13.75 | | | |
| m | 3.59 | | 3.59 | | | | 3.21 | | | |
| D/m | 0.66 | | 0.17 | | | | 0.55 | | | |

In this cross no yellow-eyed ♀♀ were to be expected. Their appearance can only be accounted for by assuming that one of the ♀♀ was already fertilized by a yellow-eyed ♂ from the $F_1$ before it was used for this cross. If these yellow-eyed ♀♀ are omitted, the result is quite up to expectation.

*d*. Mosaic ∽ flesh-coloured.

$d_1$. ♀ mosaic × ♂ flesh-coloured. The $F_1$ was black-eyed, as was expected, viz. 39 ♀♀ and 25 ♂♂. In addition there occurred one mosaic-eyed ♂, probably due to non-disjunction. The black-eyed individuals were bred together and yielded:

TABLE 51. $F_2$ OF MOSAIC × FLESH-COLOURED

| number | black<br>♀♀   ♂♂ | | mosaic<br>♀♀   ♂♂ | | red<br>♀♀   ♂♂ | | flesh-col.<br>♀♀   ♂♂ | | total<br>♀♀   ♂♂ | |
|---|---|---|---|---|---|---|---|---|---|---|
| 368 A | 21 | 21 | 1 | 2 | 5 | 6 | 6 | 11 | 33 | 40 |
| | 42 | | 14 | | | | 17 | | | |
| theor. 9 : 3 : 4 | 41.05 | | 13.70 | | | | 18.25 | | | |
| m | 4.24 | | 3.34 | | | | 3.69 | | | |
| D/m | 0.22 | | 0.09 | | | | 0.34 | | | |

According to the scheme FfGgHH × FfGgHo a ratio 9 black : 3 (mosaic and red) : 4 flesh-coloured should arise. This has indeed proved to be the case.

From the back-cross of black-eyed ♀♀ from the $F_1$ with flesh-coloured ♂♂ a 1 : 1 ratio was to be expected between black and flesh-coloured.

TABLE 52. BACK-CROSS: BLACK $F_1$ × FLESH-COLOURED

| number | black | | mosaic | | flesh-col. | | total | |
|---|---|---|---|---|---|---|---|---|
| | ♀♀ | ♂♂ | ♀♀ | ♂♂ | ♀♀ | ♂♂ | ♀♀ | ♂♂ |
| 353 | 17 | 22 | | 1 | 18 | 13 | 35 | 36 |
| | 39 | | | | 31 | | | |
| theor. 1 : 1 | 35.5 | | | | 35.5 | | | |

$$m = 4.21 \qquad D/m = 0.83$$

The occurrence of the mosaic-eyed ♂ may be accounted for by assuming that non-disjunction has taken place in the flesh-coloured ♂ and that the gamete without F which resulted from it, has been fused with a gamete with f of the ♀.

$d_2$. ♀ flesh-coloured × ♂ mosaic. The ♀ in this cross had the formula FFgghh, on account of which black-eyed ♀♀ and yellow-eyed ♂♂ in the ratio 1 : 1 might be expected in the $F_1$.

TABLE 53. $F_1$ OF FLESH-COLOURED × MOSAIC

| number | black | | red | | yellow | | total | |
|---|---|---|---|---|---|---|---|---|
| | ♀♀ | ♂♂ | ♀♀ | ♂♂ | ♀♀ | ♂♂ | ♀♀ | ♂♂ |
| 385 | 19 | | 3 | | | 27 | 22 | 27 |

The occurrence of red-eyed ♀♀ may be accounted for by non-disjunction, it may, however, also be due to strange intruders.

The cross of black-eyed ♀♀ × yellow-eyed ♂♂ from the $F_1$ gave only a small number of progeny. Owing to this fact no conclusion can be drawn. For the sake of completeness the findings have been given in the subjoined table 54.

TABLE 54. $F_2$ OF FLESH-COLOURED × MOSAIC

| number | black | | red | | yellow | | flesh-col. | | total | |
|---|---|---|---|---|---|---|---|---|---|---|
| | ♀♀ | ♂♂ | ♀♀ | ♂♂ | ♀♀ | ♂♂ | ♀♀ | ♂♂ | ♀♀ | ♂♂ |
| 385 A | 4 | 5 | 2 | | 5 | 4 | 5 | 4 | 16 | 13 |

Expectation 27 : 9 : 12 : 16.

Yellow-eyed ♂♂ from the $F_1$ back-crossed with mosaic-eyed ♀♀ yielded a progeny, consisting of 36 individuals.

TABLE 55. BACK-CROSS: YELLOW $F_1$ × MOSAIC

| number | black | | mosaic | | red | | total | |
|---|---|---|---|---|---|---|---|---|
| | ♀♀ | ♂♂ | ♀♀ | ♂♂ | ♀♀ | ♂♂ | ♀♀ | ♂♂ |
| 385 B | 11 | 10 | 6 | 3 | 2 | 4 | 19 | 17 |

A ratio 1 black : 1 (mosaic + red) was expected. The figures found are in accordance with this; m = 3, D/m = 1.00.

Likewise black-eyed ♀♀ from the $F_1$ were back-crossed with mosaic-eyed ♂♂.

TABLE 56. BACK-CROSS: BLACK $F_1$ × MOSAIC

| number | black | | mosaic | | red | | yellow | | flesh-col. | | total | |
|---|---|---|---|---|---|---|---|---|---|---|---|---|
| | ♀♀ | ♂♂ | ♀♀ | ♂♂ | ♀♀ | ♂♂ | ♀♀ | ♂♂ | ♀♀ | ♂♂ | ♀♀ | ♂♂ |
| 385 D | 30 | 9 | 11 | 7 | 9 | 7 | 1 | 21 | 1 | 8 | 52 | 52 |
| | 39 | | 34 | | | | 29 | | | | | |
| theor. | | | | | | | | | | | | |
| 3 : 3 : 2 | 38.25 | | 38.25 | | | | 25.50 | | | | | |
| m | 4.89 | | 4.89 | | | | 4.37 | | | | | |
| D/m | 0.15 | | 0.87 | | | | 0.80 | | | | | |

Here again the phenomenon discussed above, occurs. With reference to the above discussion the individuals recorded as yellow and flesh-coloured are added up and considered as yellow-eyed in this cross. The findings tally satisfactorily with the 3 : 3 : 2 ratio expected. In this cross there also occur a few exceptional cases in the yellow-eyed ♀♀, which may be accounted for by non-disjunction.

Yellow-eyed ♂♂ from the $F_1$ were back-crossed with flesh-coloured ♀♀. From this cross a ratio 1 yellow : 1 flesh-coloured is to be expected.

TABLE 57. FLESH-COLOURED × YELLOW $F_1$

| number | red | | yellow | | flesh-col. | | total | |
|---|---|---|---|---|---|---|---|---|
| | ♀♀ | ♂♂ | ♀♀ | ♂♂ | ♀♀ | ♂♂ | ♀♀ | ♂♂ |
| 385 C | 2 | 1 | 10 | 19 | 29 | 21 | 41 | 41 |

I cannot account for the occurrence of the red-eyed individuals. With regard to the difference in number of yellow-eyed and flesh-coloured individuals, I do not deem it impossible, may be also owing to the fact that ffGGhh individuals have arisen, that a number of yellow-eyed individuals have been classed with the flesh-coloured ones.

The black-eyed ♀♀ from the $F_1$ were likewise back-crossed with flesh-coloured ♂♂. As in this case a cross was concerned between individuals of the formulae FfGgHh and FFggho a ratio 1 black : 1 yellow : 2 flesh-coloured might be expected here.

TABLE 58. BACK-CROSS: BLACK $F_1$ × FLESH-COLOURED

| number | black | | yellow | | flesh-col. | | total | |
|---|---|---|---|---|---|---|---|---|
| | ♀♀ | ♂♂ | ♀♀ | ♂♂ | ♀♀ | ♂♂ | ♀♀ | ♂♂ |
| 385 E | 27 | 8 | 12 | 10 | 31 | 28 | 70 | 46 |
| | 35 | | 22 | | 59 | | | |
| theor. 1 : 1 : 2 | 29 | | 29 | | 58 | | | |
| m | | 4.66 | | 4.66 | | 5.37 | | |
| D/m | | 1.29 | | 1.50 | | 0.19 | | |

Accordingly this result is also up to the expectation.

In the numerous crosses which have been made, some show a result which is for some reason less satisfactory; the great majority, however, agree perfectly with the hypotheses given. It is a point of further investigation which of the two hypotheses should be favoured.

# CHAPTER IV

## A CASE OF SOMATIC MOSAICISM

### § 1. *Introduction*

*a.* Literature. Several cases of somatic mosaicism of animals are known. HYDE and POWELL (44, 1916) describe 3 mosaics in *Drosophila melanogaster*. From the mating of a blood ♀ and eosin ♂ a ♀ arose, the right eye of which was typically blood, the left eosin, as it was in an eosin ♂. After fertilization the genes w$^e$ (eosin) and w$^b$ (blood) were separated through a disturbance in the mitosis. The individual was sterile. A bilateral gynandromorph with left red eye, long wing and right white eye, short wing arose from a mating of red, truncate ♀ with white, long-winged ♂. One side of the body probably received 1 X chromosome, the other 2 X. The third case is the one in which probably owing to a separation of the sex-chromosomes in the cross of a white-eyed, long-winged ♀ with red-eyed, truncate ♂ a mosaic individual arose, which was left white, truncate, right red, long.

PATTERSON (60, 1929) X-rayed eggs of *Drosophila*, originating from the cross red × white. From some of these eggs developed imagines with white ommatids in the otherwise red eye. He accounts for this either by gene-mutation or by rupture of the chromosome. Something similar he found in individuals, arisen from X-rayed eggs, originated from the cross yellow, white, type × gray, eosin, singed.

MOHR (53, 1923) likewise found a mosaic individual with *Drosophila*. He also showed cytologically, that half the individual was haplo IV and consequently minute, the other half diplo IV, i.e. wild type.

PANSHIN (59, 1935) describes a case in *Drosophila*, in which two different allels of the lozenge-gen have arisen simultaneously. The result was that the left eye was lz$^{sp}$ (lozenge-strong) the right eye lz$^{wa}$ (lozenge-weak). In the mutation both somatic tissue and tissue of the sexual organs was involved which appeared from the offspring.

STURTEVANT (77, 1921) found two cases in *Drosophila simulans*, which he explained by somatic mutation. One case was a ♂ with white eyes, in which small yellow spots, the other was also a ♂, of which the right eye was wild type, the left partly wild type partly white.

ANNA R. WHITING (82, 1933) mentions that in *Habrobracon* „virgin or mated females, heterozygous for one or more factors for eye colour occasionally produce sons with mosaic eyes, apparently due to the fact that the second polar nucleus may function in development at the same time as the egg nucleus". According to her a ♂ with black ivory eyes was likewise mosaic in the gonads.

A peculiar fact occurred in an individual that was mosaic for ivory (o$^1$C) and cantaloup (Oc), because „the separating line between these two colours is black, apparently due to a diffusion from the non-ivory cantaloup (Oc) region into the non cantaloup-ivory (o$^1$C) region, so that the double dominant condition (black) is established phenotypically. This establishment is thus physiologically although not genetically".

BREITENBECHER (10, 1932) found in a homozygous strain of *Bruchus* 31 ♀♀ with elytra which were coloured differently, i.e. the left one black the right red, etc. On back-crossing with the recessive form no mosaic progeny arose. He assumes that somatic mutations occurred from recessive to dominant in one of the two homologous chromosomes at the moment that the formation of the elytra was differentiated.

Also in higher animals somatic mosaicism occurs.

CASTLE (12, 1922) obtained in rats from the mating of an albino (ccPP) ♀ with a pink-eyed and yellow-coated (CCpp) ♂ a three-coloured ♂ : gray-hooded with yellow dots in the gray part. The animal was mated several times from which it appeared that the three-coloured state was not inherited. CASTLE supposes that the chromosome containing the genes c and P, has disappeared during a division, in consequence of which cells were formed of the constitution Cp and CcPp (gray-hooded).

FISHER (30, 1930) made experiments on mice, and obtained a white ♀ with little black patches on both sides of the body and a chocolate patch between right eye and ear. Possibly the gene B(black) has got lost during the development.

BITTNER (9, 1932) also found in mice in the progeny of a mating between dilute brown ♀ and brown albino ♂ a mosaic ♂ being brownish on its back and dilute brown on the ventral side. He accounted for this as being a result of non-disjunction or of somatic mutation.

FELDMAN (28, 1935) thinks that in the case of mosaic eye found by him in the mouse the dominant gene has disappeared during the development.

DUNN (21, 1934) found a black mouse showing several tan spots. He considers this a result of mutation of the gene $C^+$ to $c^-$, i.e. wild type to albino. The cell in which the mutation arose probably also participated in the formation of the gonads.

PINCUS (62, 1929) too found some cases in the mouse the origin of which he explains through the loss of the dominant gene B (black).

WRIGHT and EATON (87, 1926) mention 7 cases of mosaicism in *Cavia cobaya*, 5 of which are important. One of these animals had two separated red agouti ($Cc^d$) spots — for the rest it was yellow agouti ($c^dc^d$). They assume that in a cell a mutation from $c^d$ to C occurred and from this cell both soma and germ-epithelium developed. The ratio of the gametes C : $c^d$ was as 79 : 149.

In another special case an allel A (agouti) must either have mutated to an unknown dominant allel or they must both have got lost. This ♀ had a distinct black spot in her coat in front of the right ear. The other 3 cases, two of which are intense-dilute and one intense-brownpale brown may according to them be explained by mutation or deficiency.

CASTLE (quoted after FISHER 30, 1930) had a ♂ rabbit, heterozygous for the dilution factor d, converting black into blue. This ♂ had a large blue patch on his shoulder, but otherwise behaved in his matings as heterozygous Dd. It is not sure whether this is also a case of mutation or of deficiency.

CREW and LAMY (14, 1935) found 17 cases of mosaic body colour in *Melopsittacus undulatus*. They also assume that the autosome was eliminated together with the dominant gene. In this way they also explain among other things the origin of half-siders. In this special case the chromosome would already have got lost in the first segmentation division.

*b.* Description. The mating between a red-black mosaic-eyed ♀ and a flesh-coloured ♂ gave black-eyed males and females in the $F_1$. Backcrossing these black-eyed ♀♀ to flesh-coloured ♂♂ yielded an $F_2$, consisting of 17 ♀♀ + 22 ♂♂ black-eyed and 18 ♀♀ + 13 ♂♂ with flesh-coloured eyes and one mosaic-eyed ♂.

Both eyes of this individual were two-coloured. The left was dorsally black, ventrally flesh-coloured, the right dorsally flesh-coloured, ventrally black. In both eyes the dividing line was about in the middle. For the rest the individual was perfectly normal.

## § 2. *Crosses*

The ♂ in question was mated to a flesh-coloured ♀. This cross yielded a heterogeneous $F_1$, consisting of 7 ♀♀ + 6 ♂♂ black-eyed and 27 ♀♀ + 25 ♂♂ flesh-coloured.

With the individuals from the $F_1$ the following crosses were made.

TABLE 59. CROSSES OF $F_1$ BEETLES

*a.* black × black

| number | black ♀♀ | black ♂♂ | red ♀♀ | red ♂♂ | flesh-col. ♀♀ | flesh-col. ♂♂ | total ♀♀ | total ♂♂ |
|---|---|---|---|---|---|---|---|---|
| 380 A | 10 | 17 | 1 | 2 | 10 | 8 | 21 | 27 |

*b.* flesh-coloured × flesh-coloured

| number | black ♀♀ | black ♂♂ | red ♀♀ | red ♂♂ | flesh-col. ♀♀ | flesh-col. ♂♂ | total ♀♀ | total ♂♂ |
|---|---|---|---|---|---|---|---|---|
| 380 B | | 1 | | | 16 | 28 | 16 | 29 |

*c.* flesh-coloured × black

| number | black ♀♀ | black ♂♂ | red ♀♀ | red ♂♂ | flesh-col. ♀♀ | flesh-col. ♂♂ | total ♀♀ | total ♂♂ |
|---|---|---|---|---|---|---|---|---|
| 380 E | 1 | 1 | 1 | | | 2 | 2 | 3 |

*d.* black × flesh-coloured

| number | black ♀♀ | black ♂♂ | red ♀♀ | red ♂♂ | flesh-col. ♀♀ | flesh-col. ♂♂ | total ♀♀ | total ♂♂ |
|---|---|---|---|---|---|---|---|---|
| 380 D | 15 | 9 | 4 | 4 | 13 | 14 | 32 | 27 |

From these results it appears that the mosaic character is not hereditary, seeing it does not occur either in the $F_1$ or in the $F_2$. Accordingly the ♂ was not genotypically but somatically mosaic.

It is evident that the ratio 13 black : 52 flesh-coloured in the $F_1$ is not a 1 : 1 ratio, for m = 4.03 and D/m = 4.84 for this ratio. To account for this the origin of the ♂ should be reverted to. I started with the cross red-black mosaic-eyed ♀ × flesh-coloured ♂, in the formula : ffGGHH × FFggHo. The $F_1$ of this cross was therefore black-eyed FfGgHH and FfGgHo. Females of the formula FfGgHH were back-crossed to flesh-coloured ♂♂ of the formula FFggHo. From this a ratio 1 black : 1 flesh-coloured was to be expected.

My opinion is that in this cross the ♂ in question originated as FfGgHo (why it should be Ff will be discussed later) and that during the development of the individual a change has taken place, which eliminated the gene G. As to the nature of this change there are two possibilities in this case: 1. G has mutated to g, 2. deficiency has occurred.

This may be represented by the following formulae: 1. FfGgHo → FfggHo, 2. FfGgHo → FfgoHo.

These cases are different with respect to the consequences they may have regarding the course of inheritance. For instance it is not to be expected that as a result of the mutation mentioned sub 1 secondary lethality will occur, whereas with deficiency lethality often occurs. In the case of deficiency there are two possibilities: *a*. the whole chromosome containing the gene G has been eliminated; *b*. only part of the chromosome has been eliminated.

In the case mentioned sub 2*a* lethality in the gametes is probable. Half the number of gametes formed will be a chromosome short, and as a result it is very likely to perish. This needs not be so in the case mentioned sub *b*, though it is of course possible that the vanished part is very large or contains genes essential for life, so that lethality nevertheless occurs. If this is not the case lethality may be expected, when two gametes both of imperfect deficiency are united.

Lethality in the gametes would be perceptible in the $F_1$ when a ♀ was concerned as in that case only a small number of progeny would arise. As a ♂ is concerned in this case nothing can be concluded from the number of $F_1$ individuals, because in spite of the dying off of a number of gametes yet enough of them will be formed to fertilize all egg-cells formed by the normal flesh-coloured ♀. If there is lethality in the gametes there will arise in the $F_1$ only individuals with a normal set of chromosomes. This possibility cannot be distinguished from the locus-mutation, because also in that case there will arise individuals with a complete set of chromosomes in the $F_1$.

It is different in the case of deficiency of a piece of chromosome that does not cause lethality of the gametes. Then there will arise in the $F_1$ flesh-coloured individuals, to be represented by the formula go. Only in the offspring of such an individual lethality may occur, because then there will arise zygotes from the fusion of two chromosome-complexes with deficiencies. The absence of part of the two homologous chromosomes will be capable of preventing such a zygote from developing.

In the mating together of flesh-coloured $F_1$ individuals in the cross indicated above, there originated from 2 ♀♀ and 2 ♂♂ an offspring of 56 larvae. This is indeed a very small number and it is possible that this is due to deficiency. Yet I will observe here that at the time the fertility, may be as a result of the care being technically imperfect, was on the whole not very great, so that from the slight number no conclusion can be drawn.

The other crosses made with $F_1$ individuals cannot teach us anything in this respect, because no fusion of two possibly deficient gametes could occur in them. Yet, because red-eyed individuals arose in them, these crosses convinced us of the fact that the mosaic ♂ must have been heterozygous Ff. In addition it is possible that in the matings of flesh-coloured and black-eyed $F_1$ individuals a black-eyed offspring of the formula FFGoHH or red-eyed ffGoHH individuals have arisen. Mating together two of such black-eyed or red-eyed individuals must yield either black-eyed progeny only or red-eyed ones only. For in the various combinations possible GG will be black-eyed, like Go, but oo will die off. As secondary phenomenon such a cross will have to show a slight fertility.

Four of such crosses have been made, table 60.

To be sure not much offspring arose from any of these matings, but this cannot be due to a dying off of the oo combinations. For in all 4 cases flesh-coloured individuals also arose, which indicates that all individuals used in these crosses were Gg, table 61.

This renders the probability of deficiency smaller, but cytological examination will only be capable of solving this problem definitely.

TABLE 60. CROSSES OF F$_2$ BEETLES

| number | eye colour | | number of larvae |
| | mother | father | |
|---|---|---|---|
| 380 B | red | red | 21 |
| C | red | red | 52 |
| D | black | black | 33 |
| I | red | red | 11 |

TABLE 61. CROSSES OF F$_2$ BEETLES

| number | black | | red | | flesh-col. | | total | |
| | ♀♀ | ♂♂ | ♀♀ | ♂♂ | ♀♀ | ♂♂ | ♀♀ | ♂♂ |
|---|---|---|---|---|---|---|---|---|
| 380 B | 1 ? | | 11 | 3 | 2 | 1 | 14 | 4 |
| C | | | 6 | 16 | 6 | 4 | 12 | 20 |
| D | 9 | 10 | | | 5 | 4 | 14 | 14 |
| I | | 2 | 3 | 3 | 2 | 1 | 5 | 6 |

Finally I will just discuss the question at what point of time this change, be it mutation or deficiency, has occurred.

Such a change can occur without its becoming perceptible in the sexual organs. Various of such cases are mentioned in literature. It is, however, also possible that the change affects the sexual organs. In the first case a ♂ that has remained Gg in his sexual organs would have to yield a ratio 1 black : 1 flesh-coloured in the F$_1$ on being mated to a gg ♀. In the second case on the other hand a disturbance of this 1 : 1 ratio is to be expected to the advantage of flesh-coloured. From the F$_1$ it is clear that the latter was the case. As this was a cross having the character of a back-cross, these figures give at the same time the ratio of the gametes in the male. The ratio of the gametes G : g has therefore been 1 : 4. From this it follows that in the case of mutation (eventually deficiency without lethal results for the gametes) about 5/12 of the sexual organs have genetically been Gg, about 7/12 gg. In the case of lethality in the gametes owing to deficiency of the whole chromosome 1/4 will have been Gg, 3/4 gg.

At any rate it may be concluded from the fact that both the eyes and the sexual organs have been involved in the change, that it must have occurred in a rather early stage of development.

# CHAPTER V

## LINKAGE

As already stated in the general introduction, FERWERDA has shown that the genes B (V-groove) and g (flesh-coloured eyes) lie in the same chromosome. Concerning the character V-groove I will give some parts of the description of FERWERDA: (p. 61, 62) „Bei genauer Betrachtung des Kopfes von der Dorsalseite zeigt es sich, dass die Struktur des Chitinpanzers sehr abnorm ist. Das Integument ist hier so schwach, dass man ohne jede Mühe mit einer stumpfen Nadel hindurchstecken kann, was bei einem normalen Käfer nie gelingt. Die Oberfläche des Panzers ist sehr wulstig, hier und da finden sich Stellen, wo sich nahezu kein verhärtetes Chitin gebildet hat, unregelmässige Vertiefungen im Panzer deuten diese Stellen an." (P. 62, 63) „Immer findet man eine scharf umrissene mediane, tiefe V-förmige Grube im Chitin des Schädels, ungefähr zwischen den Augen. Die Spitze des V liegt ein wenig kaudal vom kaudalen Augenrand; die Spitzen der Beine des V liegen ebenso weit rostral, wie der rostrale Augenrand." (p. 63). „Mit dieser Grube geht oft das Vorkommen hornartiger Fortsätze auf dem Chitinrand dorsal vom Auge zusammen." (p. 64). „Fast immer weisen die Käfer dieses Typus Antennenmissbildungen auf.... Auch die allgemeine Form des Kopfes und des Auges und weiter die Struktur des Auges zeigen noch einige interessante Besonderheiten." (p. 66). „Das Auge kann verschiedene Grade der Reduktion aufweisen.... Die Fazetten haben in Flachenansicht nicht eine regelmässige sechseckige, sondern eine ziemlich unregelmässige 4-6 eckige Form. Auch in der Anordnung der Fazetten gibt es nicht die Regelmässigkeit, welche sich beim normalen Auge findet."

This may suffice to give an idea of the character.

FERWERDA has not been able to ascertain the degree of linkage between the two genes because the crosses he had made were unsuitable for calculating cross-over values.

Therefore it seemed worth while to make a number of crosses in order to determine the degree of linkage and there was also an opportunity to trace whether the form BBgg is indeed lethal. FERWERDA says speaking of the cross Bbgg × Bbgg (p. 74): „Theoretisch ist also zu erwarten dass 25% der $F_2$ Individuen die genotypische Konstitution BBgg haben. Diese Individuen halte ich für nicht lebensfähig. Mit anderen Worten: es wird angenommen, dass der Faktor B eine Lethalwirkung ausübt, wenn er samt dem mit ihm gekoppeltem Augenfarbenfaktor g in homozygotem Zustand vorkommt."

All crosses have been made with individuals with flesh-coloured eyes and V-groove. Seeing that flesh-coloured is recessive, it is certain that the beetles used were homozygous gg. Some, however, appeared to be heterozygous for the factor B, which, considering the fact that they were homozygous gg, is of no further interest.

The $F_1$ of the reciprocal crosses between flesh-coloured with V-groove (ggBB) and black, normal (GGbb) consisted partly of individuals with V-groove, partly of individuals without, tables 62 and 63.

TABLE 62. $F_1$ OF BLACK NORMAL × FLESH-COLOURED, V-GROOVE

| number | black | | | | flesh-coloured normal | | total | |
|---|---|---|---|---|---|---|---|---|
| | V-groove | | normal | | | | | |
| | ♀♀ | ♂♂ | ♀♀ | ♂♂ | ♀♀ | ♂♂ | ♀♀ | ♂♂ |
| 365 | 20 | 24 | 31 | 27 | | | 51 | 51 |
| 374 | 55 | 52 | 10 | 7 | | 1 | 65 | 60 |
| 409 | 17 | 24 | 21 | 19 | | | 38 | 43 |
| total | 92 | 100 | 62 | 53 | | 1 | 154 | 154 |

TABLE 63. $F_1$ OF FLESH-COLOURED, V-GROOVE × BLACK, NORMAL

| number | black | | | | yellow | | | | flesh-col. | total | |
|---|---|---|---|---|---|---|---|---|---|---|---|
| | V-groove | | normal | | V-groove | | normal | | V-groove | | |
| | ♀♀ ♂♂ | | ♀♀ ♂♂ | | ♀♀ ♂♂ | | ♀♀ ♂♂ | | ♀♀ ♂♂ | ♀♀ ♂♂ | |
| 363 | 5 | 6 | 7 | 5 | | | | | | 12 | 11 |
| 408 | 21 | 13 | 23 | 21 | 11 | | 12 | | 3 | 44 | 59 |
| total | 26 | 19 | 30 | 26 | 11 | | 12 | | 3 | 56 | 70 |

The yellow-eyed individuals have arisen as a result of heterozygosis of black-eyed males.

Linkage was determined in the usual way by back-crossing with
the double recessive form; in addition, however, $F_1$ individuals
with black eyes and V-groove were intercrossed, tables 64, 65, 66.

TABLE 64. BACK-CROSSES OF $F_1$ BLACK, V-GROOVE × FLESH-
COLOURED, NORMAL

| number | non cross-overs | | | | cross-overs | | | | total | |
|---|---|---|---|---|---|---|---|---|---|---|
| | bG | | Bg | | BG | | bg | | | |
| | ♀♀ | ♂♂ | ♀♀ | ♂♂ | ♀♀ | ♂♂ | ♀♀ | ♂♂ | ♀♀ | ♂♂ |
| 361 C | 19 | 8 | 15 | 17 | 1 | 2 | 1 | 6 | 36 | 33 |
| 365 B | 3 | 2 | 1 | 2 | | | | | 4 | 4 |
| 374 B | 28 | 29 | 30 | 33 | 1 | | 1 | | 60 | 62 |
| 409 E | 44 | 32 | 49 | 45 | 10 | 10 | 2 | 1 | 105 | 88 |
| 408 A | 6 | 3 | 5 | 2 | 1 | | 1 | | 13 | 5 |
| 408 D | 7 | 17 | 8 | 17 | 1 | 1 | | 1 | 16 | 36 |
| total | 107 | 91 | 108 | 116 | 14 | 13 | 5 | 8 | 234 | 228 |
| | | 422 | | | | 40 | | | | |

TABLE 65. BACK-CROSSES OF FLESH-COLOURED, NORMAL × F₁ BLACK, V-GROOVE

| number | non cross-overs | | | | cross-overs | | | | total | |
|---|---|---|---|---|---|---|---|---|---|---|
| | bG | | Bg | | BG | | bg | | | |
| | ♀♀ | ♂♂ | ♀♀ | ♂♂ | ♀♀ | ♂♂ | ♀♀ | ♂♂ | ♀♀ | ♂♂ |
| 361 B | 19 | 17 | 13 | 11 | | 1 | 1 | 1 | 33 | 30 |
| 365 C | 10 | 1 | 27 | 13 | 2 | 1 | 8 | 8 | 47 | 23 |
| 374 A | 9 | 13 | 15 | 20 | | | 1 | | 25 | 33 |
| 408 B | 14 | 10 | 11 | 8 | | | | 3 | 25 | 21 |
| C | 5 | 1 | 1 | 5 | 1 | 1 | | | 7 | 7 |
| E | 25 | 18 | 22 | 14 | 1 | 2 | 1 | 2 | 49 | 36 |
| H | 57 | 60 | 49 | 47 | 1 | 2 | 5 | 4 | 112 | 113 |
| 409 D | 57 | 59 | 42 | 44 | 2 | 1 | 2 | 3 | 103 | 107 |
| total | 196 | 179 | 180 | 162 | 6 | 8 | 19 | 21 | 401 | 370 |
| | 717 | | | | 54 | | | | | |

TABLE 66. BLACK, V-GROOVE × BLACK, V-GROOVE

| number | BG | | Bg | | bG | | bg | | total | |
|---|---|---|---|---|---|---|---|---|---|---|
| | ♀♀ | ♂♂ | ♀♀ | ♂♂ | ♀♀ | ♂♂ | ♀♀ | ♂♂ | ♀♀ | ♂♂ |
| 361 | 16 | 18 | 8 | 6 | 8 | 7 | | 1 | 32 | 32 |
| 364 | 25 | 28 | 11 | 13 | 12 | 12 | | 3 | 48 | 56 |
| 363 A | 19 | 13 | 7 | 15 | 5 | 7 | | 1 | 31 | 36 |
| 408 G | 18 | 19 | 5 | 11 | 7 | 7 | | | 30 | 37 |
| total | 78 | 78 | 31 | 45 | 32 | 33 | | 5 | 141 | 161 |

The cross-over value computed from table 64 is $\dfrac{40}{462} \times 100 =$ 8.659%, the one computed from table 65 is $\dfrac{54}{771} \times 100 = 6,914\%$.

There is therefore not much difference in the number of cross-overs in females and in males.

The cross-over value from table 66, computed with the aid of the formula given by MAHBUB ALAM (1, 1929 and LAMPRECHT, 51, 1932), is 6.398%, which is in sufficient agreement with the above mentioned values.

As to the genotype BBgg, indicated by FERWERDA as lethal, it appeared, that it is not always perfectly lethal. I have namely succeeded in obtaining a stock pure for V-groove and flesh-coloured eyes. In two generations amounting to a total of 233 beetles not a single specimen without a V-groove has appeared; all of them must therefore be double homozygous BBgg.

Cross-breeding also convinced us of the viability of BBgg. Two crosses were made between 1 ♀ with flesh-coloured eyes and V-groove and 1 ♂ with flesh-coloured eyes without a V-groove. Of the 39 $F_1$ animals 38 had a V-groove and flesh-coloured eyes, and 1 ♂ was normal and had moreover black eyes.

I think I may conclude from this that in both crosses the ♀♀ had the formula BBgg.

# CHAPTER VI

## § 1. *Ultra-violet light*

In 1932 ELOFF (22) made experiments on *Drosophila melanogaster*, in which he determined the effect of ultra-violet light on the crossing-over value between the genes black and vestigial in the second chromosome. During these experiments also a number of larvae, pupae and beetles of *Tenebrio molitor* have been X-rayed in order to find out whether ultra-violet light is capable of bringing about mutations. For these experiments a quartzglass quicksilver lamp was made use of. The driving current was 3.5 amp. at 160 volts. The individuals which were to be treated, were placed in a porcelain dish of a diameter of 8 cms perpendicularly under the lamp, so that the whole bottom was exposed. Distance and duration were varied for the different experiments. Table 67 gives a survey of these experiments.

The reaction of the individuals to X-raying was different. One larva tried to hide under another, pupae soon started striking with their abdomen, beetles were evidently not so sensitive, in them at least a special reaction was never noticed. With the aid of a glass rod the larvae were kept apart if necessary. In the case of beetles and pupae, but for a single exception, a number of ♀♀ and ♂♂ were X-rayed at the same time. Directly after treatment the beetles were mated together, the pupae and larvae were placed in dishes into the incubator (26° C). For the rest the treatment of these individuals was the same as in the ordinary cultures. The beetles emerging from the exposed larvae and pupae were likewise mated together. As the possibility existed that recessive mutations should arise, an $F_2$ was grown of all cultures.

All cultures need not be discussed fully. The examination con-

TABLE 67. IRRADIATION WITH ULTRA-VIOLET LIGHT.

| number of experiment | duration in min. | distance in cm | individuals | number |
|---|---|---|---|---|
| Uv.  4 | 10 | 37 | young larvae | 10 |
| Uv.  2 | 10 | 37 | old larvae | 10 |
| Uv.  5 | 10 | 37 | ,, | 10 |
| Uv.  1 | 20 | 37 | ,, | 10 |
| Uv.  6 | 35 | 37 | ,, | 10 |
| Uv.  7 | 45 | 37 | ,, | 10 |
| Uv. 11 | 10 | 16 | young pupae | 4 ♀♀ + 3 ♂♂ |
| Uv.  8 | 40 | 23 | ,, | 4 |
| Uv.  3 | 10 | 37 | old pupae | 8 |
| Uv.  9 | 40 | 16 | ,, | 10 |
| Uv. 12 | 10 | 16 | young beetles | 2 ♀♀ + 3 ♂♂ |
| Uv. 15 | 28 | 16 | ,, | 3 ♀♀ + 3 ♂♂ |
| Uv. 10 | 40 | 16 | ,, | 2 ♀♀ |
| Uv. 13 | 10 | 16 | old beetles | 1 ♀ + 1 ♂ |
| Uv. 14 | 30 | 16 | ,, | 5 ♀♀ + 5 ♂♂ |

TABLE 68. OFFSPRING OF IRRADIATED INDIVIDUALS

| number of experiment | cross of ♀♀ | cross of ♂♂ | eggs | larvae | percentage | duration of irradiation |
|---|---|---|---|---|---|---|
| Uv.  4 | | | | | | 10 |
| Uv.  2 | 3 | 1 | 132 | 18 | 13.6 | 10 |
| Uv.  5 | 2 | 2 | 144 | 74 | 51.4 | 10 |
| Uv.  1 | | | 85 | 18 | 21.2 | 20 |
| Uv.  6 | 1 | 1 | 89 | 41 | 46.7 | 35 |
| Uv.  7 | 2 | 2 | 119 | 81 | 68.1 | 45 |
| Uv. 11 | 5 | 2 | 233 | 115 | 49.4 | 10 |
| Uv.  8 | 1 | 1 | 165 | 91 | 55.2 | 40 |
| Uv.  3 | 2 | 2 | 185 | 60 | 32.4 | 10 |
| Uv.  9 | 1 | 1 | 98 | 40 | 40.8 | 40 |
| Uv. 12 | 2 | 2 | 196 | 108 | 55.1 | 10 |
| Uv. 15 | 3 | 3 | 263 | 195 | 74.1 | 28 |
| Uv. 10 | 1 | 1 | 87 | 53 | 60.9 | 40 |
| Uv. 13 | 1 | 1 | 136 | 64 | 47.0 | 10 |
| Uv. 14 | 5 | 5 | 354 | 180 | 50.8 | 30 |

cerned eye- and body colour, number of joints of the antennae, number of tarsal segments of the legs, and habitus of the elytra. In 853 $F_1$ individuals and 1814 $F_2$ individuals not a single mutation was found. There did occur some elytra abnormalities, but they were already known to ARENDSEN HEIN, who ascertained that they were not hereditary. Moreover the number of these cases was not larger than their number in normal strains.

Further the intention was to discover whether the fertility of the exposed individuals had changed. It may best be ascertained by comparing the number of eggs produced and the number of larvae emerged from them. A survey of this is given in table 68.

From this table it appears that on longer exposure a greater percentage of eggs hatch. An exception to this is Uv. 10, for with respect to Uv. 15 it should show an increase in percentage, whereas in reality a decrease has occurred. It may, however, be that a longer exposure is injurious to young beetles. But seeing the number of eggs is small in this case, comparison is impossible.

It is difficult to determine the number of eggs averagely laid by one ♀, as there is a difference between a pair-mating and a mating between more ♀♀ and ♂♂. My experience is that on the whole averagely per ♀ more eggs are laid in the case of a mating of 3 or 4 ♀♀ with 3 or 4 ♂♂, than in a pair-mating.

## § 2. *Length*

ARENDSEN HEIN measured a great number of pupae and selected them according to length. He obtained 4 different genotypes respectively with an average size of $M \pm 3m = 14.93 \pm 0.06, 16.15 \pm 0.06, 16.75 \pm 0.12$ and $18.74 \pm 0.13$ mms. At the beginning of my research nothing was left of this material. Therefore new experiments were commenced, which were started with one cross of small × large. In contradistinction with ARENDSEN HEIN I measured the beetles instead of the pupae. With a view to comparing the lengths of pupae and beetles, 10 pupae were measured and next the beetles emerging from them. The average length of the pupae was 1.202 times that of the beetles, varying between 1.16 and 1.24. ARENDSEN HEIN's pupae-lengths therefore correspond with the beetle-lengths 12.42, 13.435, 13.93 and 15.59 mms. By applying selection to the progeny of the above cross a strain has now been obtained, which had the average sizes M $\pm$ 3 m $=$ 14.02 $\pm$ 0.294, 14.318 $\pm$ 1.005 and 13.704 $\pm$ 0.264 mms in 3 successive generations and an other strain for which in 4 successive generations the average lengths M $\pm$ 3 m $=$ 16.24 $\pm$ 0.237, 16.13 $\pm$ 0.321, 16.24 $\pm$ 0.147 and 16.262 $\pm$ 0.195 mms were found. The difference between these two strains is statistically significant. Reciprocal crosses between this small and this large strain yielded an $F_1$ which in the cross small $\times$ large consisted of 197 individuals of an average length of 14.967 $\pm$ 0.135 mms, in the cross large $\times$ small of 173 individuals of an average length of 15.124 $\pm$ 0.132 mms. The $F_1$ therefore appears to be intermediary. These $F_1$ beetles have been mated together and back-crosses have been made of the $F_1$ with small beetles. The individuals obtained are, however, only in the larva-stage.

As contrasted with ARENDSEN HEIN I have found that the ♂♂ are larger than the ♀♀. The average length of 1169 ♂♂ amounted to 15.752 $\pm$ 0.084 mms, that of 1260 ♀♀ to 15.252 $\pm$ 0.099 mms. It should be observed that the length was always measured before the sex was determined.

§ 3. *Experiments on the possibility of mating 1 ♀ successively with more ♂♂*

After removal of the ♂ the ♀ still continues producing eggs, though in a less degree. Accordingly it is in the first place important to know how long these eggs of a mated ♀ are still being fertilized after removal of the ♂, because in a second mating no influence of the first ♂ must of course be perceptible. The experiment was conducted as follows: ♀♀ and ♂♂ were brought together for some days. The eggs laid were gathered. When it appeared that larvae emerged from these eggs, the ♂♂ were removed (Dec. 17th) and the eggs were collected in special dishes, table 69.

TABLE 69. FERTILITY OF EGGS AFTER REMOVING THE ♂♂

|        | Dec. 17 | Dec. 24 | Dec. 28 | Jan. 3 | Jan. 7 | Jan. 10 | Jan. 14 | Jan. 17 | Jan. 21 | Jan. 28 | Febr. 4 | Febr. 11 |
|--------|---------|---------|---------|--------|--------|---------|---------|---------|---------|---------|---------|----------|
| eggs   | 32      | 19      | 24      | 19     | 17     | 4       | 7       | 5       | 2       | 4       | 5       | 7        |
| larvae | 21      | 13      | 15      | 14     | 13     | 1       | 1       | 1       | 0       | 0       | 0       | 0        |

From the 5 eggs laid on January 17th, i.e. a month after removal of the ♂♂ 1 larva developed. At any rate it is possible to gather a fair amount of eggs up to a fortnight after removal of a ♂, after that the laying of eggs is greatly diminished.

It was also an important question how long it had to be waited before ♀♀ that had been mated, could be mated again, after the first ♂ had been removed. It seems as if the above experiment already solves this problem, but there is one objection. In this experiment the possibility could not be taken into account that the sperm of the first ♂ becomes less mobile after some time and may consequently be entirely ousted by that of the second ♂, a fact that may be compared with certation in plants. That is why the following experiment has been made : on April 16th 1 ♀ and 1 ♂ were mated, both having flesh-coloured eyes. The eggs produced were gathered. On April 25th the ♂ was removed, and immediately replaced by a black-eyed one. After that the eggs were again gathered. On May 6th this ♂ was removed and immediately replaced by a black-eyed melanistic individual, table 70.

TABLE 70. PROGENY OF A ♀, MATED THREE TIMES

| | black | | | | yellow | | | |
| | mel. | | d. br. | | mel. | | d. br. | |
| | ♀♀ | ♂♂ | ♀♀ | ♂♂ | ♀♀ | ♂♂ | ♀♀ | ♂♂ |
|---|---|---|---|---|---|---|---|---|
| April 29 | | | 1 | | | | | 2 |
| May 2 | | | 2 | | | | | 3 |
| 6 | | | 6 | | | | | 5 |
| 10 | | | 2 | | | | | |
| 13 | 2 | | 1 | | 1 | | | |

From the table it appears that in the eggs gathered on April 29th the influence of the first ♂ could already no more be traced, on May 13th the melanistic ♂ began to predominate.

In a second experiment on April 16th 1 ♀ was mated with 1 ♂ both having flesh-coloured eyes. On April 25th the ♂ was removed, and on April 29th replaced by a black-eyed ♂. Eggs were gathered on April 25th and 29th, which were to give flesh-coloured beetles only and further on May 6th, 10 th, 13th, and 16th, table 71.

TABLE 71. PROGENY OF A ♀ MATED AGAIN 4 DAYS AFTER REMOVAL OF
THE FIRST ♂

|  | flesh-col. | | black | | yellow | |
|---|---|---|---|---|---|---|
|  | ♀♀ | ♂♂ | ♀♀ | ♂♂ | ♀♀ | ♂♂ |
| April 25, 29 | 4 | 6 |  |  |  |  |
| May      6 | 1 |  | 3 | 1 |  |  |
| 10 |  |  | 1 | 2 |  |  |
| 13 |  |  | 6 | 1 |  | 1 |
| 16 |  |  | 3 | 1 |  | 2 |

From the table it appears that about a fortnight after removal of the ♂ its influence is no more to be noticed.

Finally in a third experiment a black-eyed melanistic ♂ was added 8 days after removal of the first ♂, table 72. From the eggs gathered on April 29th, and on May 2nd and 7th, only beetles with flesh-coloured eyes originated, as might be expected. On May 13th the influence of the ♂ removed on April 29th had disappeared.

TABLE 72. PROGENY OF A ♀ MATED AGAIN 8 DAYS AFTER REMOVAL OF
THE FIRST ♂

|  | flesh-col. | | black, mel. | | yellow, mel. | |
|---|---|---|---|---|---|---|
|  | ♀♀ | ♂♂ | ♀♀ | ♂♂ | ♀♀ | ♂♂ |
| April 29, May 2, 7 | 16 | 22 |  |  |  |  |
| May      10 |  | 3 | 6 | 3 |  | 1 |
| 13 |  |  | 18 | 14 |  | 5 |
| 16 |  |  | 11 | 4 |  | 6 |
| 20 |  |  | 3 | 1 |  | 5 |

From these experiments it appears that a ♀ can be used for a new mating, if good care is taken that the eggs which are laid during the first fortnight after removal of the first ♂, are also removed.

SUMMARY

1. The difference in genes existing between the dark brown and the melanistic variety of *Tenebrio molitor* is expressed in a difference between the ferments which play a part in the formation of the

pigment of these two varieties. The chromogenes from which these pigments arise, are most probably identical.

2. Under the influence of the ferment of the dark brown variety there is also formed from pyrocatechin a pigment different from the one formed under the influence of the ferment of the melanistic type.

3. Larvae and beetles of the same variety have the same ferment.

4. A hypothesis has been given concerning the material differences between the known eye colour types based upon the genetic differences.

5. The triple recessive eye colour variety ffgghh is flesh-coloured.

6. The character „mosaic eye" is inconstant. On mating mosaic-eyed individuals together there always arise mosaic-eyed offspring in addition to red-eyed ones. To account for this fact two hypotheses have been given, both in accordance with the results of the cross-breeding. In the one a labile gene f is adopted, mutating to F, in the second a translocation of a piece of chromosome with the gene F to the chromosome with the gene f is assumed. This translocated piece is supposed to disappear often during the divisions.

7. In a cross of mosaic with flesh-coloured a ♂ arose of which the eyes were partly black partly flesh-coloured. This has been explained by the disappearance or the mutation of the dominant allel G, owing to which the recessive allel gets an opportunity to be effective.

8. The linkage of the genes B (V-shaped head-groove) and g (flesh-coloured eyes) observed by FERWERDA had been determined.

9. Larvae, pupae and beetles have been exposed to ultra-violet light. This has not given rise to mutations. Ultra-violet light seems to have a favourable effect on the hatching of the eggs.

10. Two strains of a constant length have been isolated. Reciprocal crosses between these strains gave an intermediary $F_1$.

11. The eggs of a ♀ are still fertilized 4 weeks after the removal of the ♂.

12. It is possible to cross a ♀ several times. The offspring from the eggs laid during the first fortnight after the removal of the first ♂ should not be added to the second cross.

LITERATURE

1. ALAM, M., 1929. The calculation of linkage values. Memoirs of the Department of Agriculture in India. Vol. 18.
2. ARENDSEN HEIN, S. A., 1920. Technical experiences in the breeding of *Tenebrio molitor*. Proc. Kon. Acad. Wet. Amst. Vol. 23, p. 193.
3. ARENDSEN HEIN, S. A., 1920. Studies on variation in the mealworm *Tenebrio molitor*. I. Biol. and gen. notes on *Ten. mol.* Journ. Gen. Vol. 10, p. 227.
4. ARENDSEN HEIN, S. A., 1923. Larvenarten von der Gattung *Tenebrio* und ihre Kultur (Col.). Ent. Mitt. Bd. 12, p. 121.
5. ARENDSEN HEIN, S. A., 1924. Selektionsversuche mit Prothorax und Elytravariationen bei *Tenebrio molitor*. Ent. Mitt. Bd. 13, p. 153 and 243.
6. ARENDSEN HEIN, S. A., 1924. Studies on variation in the mealworm *Tenebrio molitor*. II. Variations in tarsi and antennae. Journ. Gen. Vol. 14, p. 1.
7. ATTA, E. W. v., 1932. Dominant eye colours in *Drosophila*. Am. Nat. 66, p. 93.
8. BÉLÂR, K., 1928. Die Technik der descriptiven Cytologie. Peterfi I p. 638.
9. BITTNER, J. J., 1932. A color-mosaic in the mouse. Journ. Hered. 23.
10. BREITENBECHER, J. K., 1932. Somatic mutations and elytral mosaics of *Bruchus*. Biol. Bull. 43, p. 10.
11. BRIDGES, C. B., 1925. Elimination of chromosomes due to a mutant (Minute-n) in *Drosophila melanogaster*. Proc. Nat. Acad. Sci. 11, p. 701.
12. CASTLE, W. E., 1922. Genetic studies in rabbits and rats. part II: On a non-transmissable tricolor variation in rats. Publ. Carn. Inst. 320 p. 51.
13. CASTLE, W. E., 1929. A mosaic (intense-dilute) coat pattern in the rabbit. Journ. exp. Zoöl. 52, p. 471.
14. CREW, F. A. E. and LAMY, R., 1935. Autosomal colour mosaics in the Budgerigar. Journ. Gen. 30, p. 233.
15. DANNEEL, R., 1936. Die Färbung unserer Kaninchenrassen und ihre histogenetischen Grundlagen. Z. f. i. A. u. V. 71, S. 231.
16. DEMEREC, M., 1926. Reddish, a frequently „mutating" character in *Drosophila virilis*. Proc. Nat. Acad. Sci. 12, p. 11.
17. DEMEREC, M., 1926. Miniature-Alpha, a second frequently mutating character in *Drosophila virilis*. Ibid. p. 687.
18. DEMEREC, M., 1927. Magenta-Alpha, a third frequently mutating character in *Drosophila virilis*. Proc. Nat. Acad. Sci. 13, p. 249.
19. DEMEREC, M., 1927. The behavior of mutable genes. Verh. d. V int. Kongr. f. Vererb. I, p. 183.
20. DEMEREC, M., 1931. Behavior of two mutable genes of *Delphinium ajacis*. Journ. Gen. 24, p. 179.
21. DUNN, L. C., 1934. Analysis of a case of mosaicism in the house mouse. Journ. Gen. 29, p. 317.

22. ELOFF, G., 1932. A theoretical and experimental study on the changes in the crossing-over value, their causes and meaning. Genetica 14, p. 1.

23. EMERSON, R. A., 1913. The inheritance of a recurring somatic variegation in variegated ears of maize. Am. Nat. 48, p. 87.

24. EMERSON, R. A., 1929. The frequency of somatic mutation in variegated pericarp of maize. Genetics 14, p. 488.

25. EYSTER, W. H., 1924. A genetic analysis of variegation. Genetics 9, p. 372.

26. EYSTER, W. H., 1925. Mosaic pericarp in maize. Genetics 10, p. 179.

27. EYSTER, W. H., 1928. The mechanism of variegation. Verh. d. V intern. Kongr. f. Vererb. Bd. I, p. 666.

28. FELDMAN, H. W., 1935. A mosaic (Dark eyed intense-pink eyed dilute) coat colour of the house mouse. Journ. Gen. 30, p. 383.

29. FERWERDA, F. P., 1928. Genetische Studien am Mehlkäfer. Genetica 11, p. 1.

30. FISHER, R. A., 1930. Note on a tricolor (mosaic) mouse. Journ. Gen. 23, p. 77.

31. FREDERIKSE, A. M., 1924. Rudimentary parthenogenesis in *Tenebrio molitor* L. Journ. Gen. 14.

32. FREDERIKSE, A. M., 1926. Species crossing in the genus *Tenebrio*. Journ. Gen. 16.

33. FÜRTH, O. v. und SCHNEIDER, H., 1901. Über tierische Tyrosinasen und ihre Beziehungen zur Pigmentbildung. Hofmeisters Beitr. z. chem. Phys. und Path. Bd. 1, p. 229.

34. GLASS, H. B., 1932. A study of dominant mosaic eye-color mutants in *Drosophila*. Proc. sixth int. congr. gen. II, p. 62.

35. GLASS, H. B., 1934. A study of dominant mosaic eye colour mutants in *Drosophila melanogaster*. Journ. Gen. 28, p. 69.

36. GOLDSCHMIDT, R., Physiologische Theorie der Vererbung. Berlin 1927.

37. GOWEN, J. W. and GAY, E. H. 1933. On eversporting as a function of the 4-chromosome in *Drosophila melanogaster*. Am. Nat. 67, p. 68.

38. GREB, R. J., 1933. Effects of temperature on production of mosaics in *Habrobracon*. Biol. Bull. 65.

39. HAAN, H. DE, 1933. Inheritance of chlorophylldeficiencies. Bibliographia Genetica 10, p. 358.

40. HAECKER, V., 1918. Entwicklungsgeschichtliche Eigenschaftsanalyse (Phaenogenetik). Jena.

41. HANSON, B. and WINKLEMAN, E., 1929. Visible mutations following radium irradiation in *Drosophila melanogaster*. Journ. Hered. 20, p. 277.

42. HASEBROEK, K., 1922. Untersuchungen zum Problem des neuzeitlichen Melanismus der Schmetterlinge. Fermentforschung 5, p. 1 and 297.

43. HYDE, R. R., 1915. The origin of a new eye colour in *Drosophila repleta* and its behaviour in heredity. Am. Nat. 44, p. 185.

44. HYDE, R. R. and POWELL, H. M., 1916. Mosaics in *Drosophila ampilophila*. Genetics 1, p. 581.

45. IBSEN, H. L., 1916. Tricolor inheritance. Genetics 1, pp. 287, 367, 377.

46. IMAI, Y., 1930. Studies on yellow-inconstant, a mutating character of *Pharbitis nil*. Journ. Gen. 22, p. 191.

47. IMAI, Y. and KANNA, B., 1935. A form of *Portulaca grandiflora*, bearing creamish flowers with yellow and orange stripes. Genetica 17, p. 27.

48. KLINCKSIECK et VALETTE. Code des couleurs.

49· KÜHN, A., 1927. Die Pigmentierung von *Habrobacon juglandis*, ihre Prädetermination und ihre Vererbung durch Gene und Plasmon. Nachr. d. Ges. der Wiss. zu Göttingen.

50. KÜHN, A., CASPARI, E., PLAGGE, E., 1935. Über hormonale Genwirkungen bei *Ephestia Kühniella* ZELLER. Nachr. Ges. Wiss. Gött. 2, S. 1.

51. LAMPRECHT, H., 1932. Zur Genetik von *Phaseolus vulgaris* IV. Hereditas 17, S. 21.

52. MALINOWSKY, E., 1935. Studies on unstable characters in *Petunia*. Genetics 20, p. 342.

53. MOHR, O., 1923. A somatic mutation in the singed locus of the X-chromosome in *Drosophila melanogaster*. Hereditas 4, p. 142.

54. MORGAN, T. H. and BRIDGES, C., 1913. Dilution effects and bicolorism in certain eyecolors of *Drosophila*. Journ. of exp. Zool. 15, p. 429.

55. MORGAN, T. H., BRIDGES, C. and STURTEVANT, A. H., 1925. Genetics of *Drosophila*. Bibl. Gen. 2, p. 1.

56. MORGAN, L. V., 1929. Contributions to the genetics of *Dros. simulans* and *Dros. melanogaster*. Publ. Carn. Inst. Wash. 399, p. 1.

57. MULLER, H. J., 1930. Radiation and Genetics. Am. Nat. 64, p. 220.

58. MULLER, H. J., 1930. Types of visible variations induced by X-rays in *Drosophila*. Journ. Gen. 22, p. 299.

59. PANSHIN, I., 1935. The analysis of a bilateral mosaic mutation in *Drosophila melanogaster*. Bull. of the Inst. of Gen. Moskou, 10, p. 231.

60. PATTERSON, J. T., 1929. X-rays and somatic mutations. Journ. of Hered. 20, p. 261.

61· PATTERSON, J. T., 1932. A new type of mottled eyed *Drosophila* due to an unstable translocation. Genetics 17, p. 38.

62. PINCUS, G., 1929. A mosaic (black-brown) coat pattern in the mouse. Journ. exp. Zool. 52, p. 439.

63. PLOUGH, H. H., 1927. Black suppressor a sex-linked gene in *Drosophila* causing apparent anomalies in crossing-over in the second chromosome. Verh. d. V int. Kongr. II, p. 1193.

64. PRZIBRAM, H. und L. BRECHER, 1919. Ursachen tierischer Farbkleidung. I Vorversuche an Extrakten. Arch. f. Ent. Mech. d. Org. 45, S. 83.

65. PRZIBRAM, H., Ursachen tierischer Farbkleidung. II Die Theorie. Ibid. S. 199.

66. PRZIBRAM, H. und DEMBOWSKY, J., Ursachen tierischer Farbkleidung. III Konservierung von Tyrosinase durch Luftabschluss. Ibid., S. 260.

67· PRZIBRAM, H., DEMBOWSKY, J. und BRECHER, L., 1921. Ursachen

tierischer Farbkleidung. IV Einwirkung von Tyrosinase auf „Dopa".
Arch. f. Entw. Mech. d. Org. 48, S. 140.

68. SCHMALFUSS, H., H. BARTHMEYER, H. BRANDES, 1928. Über das Ent-
stehen von Melanin in Organismen. Z. f. i. A. u. V. 47.

69. SCHMALFUSS, H., 1928. Vererbung, Entwicklung und Chemie nebst
entwicklungschemischen Untersuchungen an Organismen. Die Na-
turwiss. 16, p. 209.

70. SCHMALFUSS, H. und H. BARTHMEIJER, 1930. Vererbungstheoretische
Betrachtungen nebst, u.s.w. Z. f. i. A. u. V. 53, p. 67.

71. SCHMALFUSS, H. und H. WERNER 1926. Chemismus der Entste-
hung von Eigenschaften. Z. f. i. A. u. V. 41, p. 285.

72. SCHMALFUSS, H., H. BARTHMEIJER, W. HINSCH, 1931. Vererbungstheoretische Betrachtungen nebst entwicklungschemischen Untersuchungen etc. Z. f. i. A. u. V. 58, p. 332.

73. SHULTZ, J., 1932. The behavior of vermilion-suppressor in mosaics.
Proc. nat. acad. sci. 18.

74. SPENCER, W. P., 1926. The occurrence of pigmented facets in white eyes
in *Drosophila melanogaster*. Am. Nat. 60, p. 282.

75. SPENCER, W. P., 1930. Mosaic-orange, an asymmetrical eye-color in
*Dros. hydei*. Journ. of exp. Zool. 56, p. 267.

76. STUBBE, H. 1933. Labile Gene. Bibl. Gen. 10, p. 299.

77. STURTEVANT, A. H., 1921. Genetic studies on *Drosophila simulans*.
Genetics 6, pp. 43, 179.

78. SURRARRER, T. C., 1935. The effect of temperature on a mottled-eye
stock of *Drosophila melanogaster*. Genetics 20, p. 357.

79. VERNE, J., 1926. Les pigments dans l'organisme animal. Paris.

80. WHITING, P. W., 1926. Heredity of two variable characters in *Habrobracon*. Genetics 11, p. 305.

81. WHITING, P. W., 1928. Mosaicism and mutation in *Habrobracon*. Biol.
Bull. 54, p. 289.

82. WHITING, ANNA R., 1933. Variegated eye-color in the parasitic wasp
*Habrobracon*. The collecting Net. 8.

83. WHITING, P. W., 1932, 1934, Mutants in *Habrobracon*. Genetics 17, p. 1
and Genetics 19 p. 268.

84. WHITING, A. R., 1934. Eye-colours in the parasitic wasp *Habrobracon*
and their behaviour in multiple recessives and in mosaics. The collecting Net 8, Journ. Gen. 29, p. 99.

85. WHITING, P. W. and A. R., 1934. A unique fraternity in *Habrobracon*.
Journ. Gen. 29, p. 311.

86. WIT, F. 1937. Contributions to the genetics of the China Aster. Genetica
19, p. 1.

87. WRIGHT, S. and EATON, O. N., 1926. Mutational mosaic coat patterns in
the guineapig. Genetics 11, p. 333.